T5-AGA-405

Technological
Planning
and
Social Futures

Technological Planning and Social Futures

Erich Jantsch

A HALSTED PRESS BOOK

JOHN WILEY & SONS
New York

658.4
J35t

Published in the U.S.A.
by Halsted Press, a Division
of John Wiley & Sons, Inc.
New York

First published 1972

Jantsch, Erich.
  Technological planning and social futures.

  "A Halsted Press book."
  Bibliography: p.
  1. Technological forecasting.   2.   Technology—
Social aspects.   I.   Title.
T174.J33   1972b   658.4   72-5443
ISBN 0-470-43997-1

© Erich Jantsch 1972

This book has been printed in Great Britain
by The Anchor Press Ltd., and bound by
Wm. Brendon & Son Ltd., both of Tiptree, Essex

For Ritva and Matti
From whom I keep learning

SEP 6 73

HUNT LIBRARY
CARNEGIE-MELLON UNIVERSITY

Acknowledgments

The author and the publishers wish to express their thanks to the following organisations for permission to reprint material which first appeared in their publications (the exact references appear at the beginning of the six thematic subdivisions of the book):

American Elsevier Publishing Company, Inc., New York: Chapters 2, 5, 10, 11, 15, and 16.

Organisation for Economic Co-operation and Development (OECD), Paris: Chapters 3 and 9.

The Editor of *Long Range Planning*, Oxford: Chapter 4.

La Recherche/Atomes, Paris: Chapter 7.

Prentice-Hall, Inc., Englewood Cliffs, New Jersey: Parts of Chapter 8.

The Institution of Chemical Engineers, London: Parts of Chapter 8.

The Editors of *The Columbia Journal of World Business*, and Columbia University, New York: Chapter 13.

The Editors of *Futures*, Guildford, and IPC Science and Technology Press Ltd., London: Chapter 14.

Also, the author and the publishers are grateful to Professor *Bertrand de Jouvenel* for kindly agreeing to write a Foreword.

Contents

Illustrations and Tables

TABLES

Foreword

THE more we recognise technological change as a major cause of social change the more we must be attentive to the formation of technological change.

This has altered a great deal from the days when a gifted mechanic convinced receptive industrialists to try a steam locomotive. Compare with the Manhattan project when an experiment in Berlin, at the utmost heights of pure science, was perceived in the U.S. as a source which could be channelled rapidly towards a concrete operation. The shattering and hideous success obtained served as an advertisement for the efficiency of an integrated process leading from scientific discoveries to practical innovations.

In this new R & D era, it has become a normal intellectual pursuit to scan scientific findings for the practical possibilities they contain and that is the spirit of technological forecasting. Erich Jantsch established his reputation by a monumental investigation of procedures used in such forecasting. Therefrom a moral concern led him to the theme of technological planning.

It is a widely prevalent superstition that we are blown into the future by the uncontrollable winds of technological change. But how can they be uncontrollable since they are of our own making?

Not only does a feasible innovation require an investment for its operation, but the achievement of feasibility absorbs much work and equipment and calls for much funding. Need I add that the further up we move from the latest stages of development to the higher sources of science the less the funding can be bound to a specific project and the more it must be viewed in terms of general orientations.

The funding agencies therefore find themselves agents of society for the guiding of technological change. The question then arises: are they good agents?

As early as 1603 Francis Bacon advanced the idea (in *Valerius Terminus*) that the diverse activities which we now embrace under the term of R & D should be supported in proportion 'to the use and tribute which they yield and render to the conditions of man's life'.

Here then a principle was laid down for an ideal allocation of resources.

Tremendous would be the intellectual difficulties involved in working down from the principle to positive statements regarding desirable allocation, and great the variety of answers given by different minds making different assumptions and using different value judgments. But however great this variety, one thing can be said with assurance: the actual distribution of resources lies far outside the field embracing such conceivable answers.

And there is no reason for the actual distribution to lie within this field, there being no social mechanisms or devices attracting R & D efforts in the service of man's condition.

Such a mechanism is at work only in the case of consumer-oriented industries and their equipment providers. And here it is affected by well-known deficiencies: neglect of the poorer consumers (hence of Third World needs) and indifference to external effects (e.g. environmental harms).

But far the greater part of R & D resources are directly oriented towards Government requirements and here we see a pull of talents towards power and prestige achievements. We find no device whereby Government requirements are tailored to the conditions of man's life. While the sovereignty of the consumer is not all that theoretical economics assume, even less is the sovereignty of the citizen such as political theory assumes.

The late twentieth century stands in direct contrast to the expectations of the late eighteenth century. Individuals depend in every way upon huge organisations, be they corporations or government agencies. There is concentration of means, of decision making, of power upon the future, of responsibility. Jantsch addresses himself to those who sit in high places and urges them to consider not only their immediate function but its place in a vast system of service to Mankind.

Bertrand de Jouvenel

1. Introduction

THE concept of corporate long-range planning initiated a process of profound change in business attitudes and general thinking. As a widespread phenomenon, this process became visible in the United States in the first half of the 1950s, and in Europe in the early 1960s. However, we have probably witnessed only a beginning so far. The consequences of consciously and rationally orienting thinking and action toward long-range futures cannot yet be fully measured out – they may be expected, not only to revolutionise the ways in which corporations operate, but also to,lead to a thorough redefinition of roles and tasks for business in a rapidly changing societal and world context.

Long-range thinking implies developing a *sense of direction*. Where a succession of short-range steps may be made *in reaction* to changes in the environment, a long-range framework requires thinking and acting *in anticipation*. Not, that a rigid yard-stick is to be cast into a hazy future – as for short-range operational steps – but direction and thrust of action have to be understood, monitored and continuously modified. This brings the question of 'good' and 'bad' to the foreground: What is a 'good' direction, and how can we obtain criteria by which to distinguish between 'good' and 'bad' action? A whole host of problems is thereby introduced to the decision making process, problems of personal and social values, of pluralism and consensus, of motivation and organisation, of decentralised initiative and co-ordination – in short, of creative participation.

In a long-range time-frame, only the short-range end of planning has to be 'frozen in' for immediate realisation. For the more distant future, *alternative options* are brought into view and kept open as long as possible. Strategic planning is not about one inescapable future, but about a multitude of possible futures (or 'futuribles', as Bertrand de Jouvenel calls them).

Hand in hand with this new sense for direction – and for the *outcome* of planning and action – goes an awareness of the *systemic nature* of the reality to which planning and action are applied. The naive belief that the optimisation of subsystems sums up to an optimisation of the overall system – with a macroautomatism, such as Adam Smith's 'hidden hand', taking care of the overall system – can no more be upheld. Also, in a long-range time-frame, restriction to narrow angles of view – looking separately at economic, social, technological, political, anthropological aspects of the future – becomes absurd. The emphasis is on systemic or *integrative* thinking, cutting across all these dimensions simultaneously. In particular, technological planning *per se* is not a viable concept within a long-range framework. In this respect, the title of the book is imprecise. Nor can the straightforward, sequential problem-solving approach, so successful when applied to purely technological tasks, be applied to planning and action in an integrative spirit. New approaches to systemic forecasting and planning are gradually emerging, but are yet in a primitive stage of development, making long-range planning an art rather than a set of rigorously pre-scribed and logical procedures.

Until recently, the dominant modes of forecasting and planning were *prediction* (of what 'will' happen) and *'linear' planning* along given tracks and within a given framework, assuming that the forces that have acted in the recent past will continue to act in the near future. It may be considered fortunate that these types of forecasting and planning, and the considerable sophistication developed in them, continue to be useful in the full long-range planning context – but in the specific role of *operational planning* now linking up with strategic and normative planning (the 'can' and the 'ought to').

The discovery – or, better, rediscovery – that, to a large extent, man holds the options in his hands to shape his own future, and may aim at alternate 'futuribles' (possible futures), electrified many people in the 1960s. In Europe, *Bertrand de Jouvenel* and *Dennis Gabor* became the first spiritual leaders for a widespread interest in the future, in America *Herman Kahn*, *Olaf Helmer* and *Daniel Bell*.

The further step of interpreting and understanding our situation in terms of complex system dynamics, is more difficult to make. We are grappling with it now. German 'futurology' and other dilettante and playful modes of futurism, leading to the formation of parochial cliques devoting themselves to a new mysticism, provide surrogates for the more simple-minded. However, they are also dangerous, as thinking about the future without explicit links to action in the present – how to get from here to an imagined future – creates discouragement and even dis-enchantment with the future. 'The futuristic fog-machine pollutes clear thinking about the present' – with these words, students recently marched out of a conference of loose and unrealistic futuristic speculation.

To go back to a mere 'activist' or short-range attitude is also a big temptation for the professional planner of a conventional type. He has pushed operational planning in the belief it contributes to the reduction of uncertainty and complexity – only to find now that the full long-range planning process is geared to *increase uncertainty* (by bringing as many options as possible into play) as well as *complexity* (by attempting to grasp the systemic context of a particular plan).

Most difficult to accept, however, may be the *cybernetic* approach which is inherent in full-scale long-range planning: Through our actions, we influence the environment in which we live, and which acts back on our ways of life. We cannot predict this feedback with any certainty if we go to a long-range time-frame. Thus it is necessary to keep the options open, to understand the dynamics of the social systems which we are part of, to comprehend our means of influencing system dynamics, and to bring them into play in a flexible way. This type of planning is not concerned about *how* to get from point *A* to point *B* – or only at the operational level, dealing with the short range – but about *what* would be a good point *B* to choose, which strategy would bring us there in a 'good' way, and *where* our systems of human living would be moving in dependence of our choice.

If long-range planning is so different and so difficult, why, then, are governments and corporations trying to incorporate it into their thinking? There are two main reasons: One has to do with the pace at which change is taking place in a planned or unplanned way, and in particular, techno- logical change. While it may be possible to stumble along in darkness in small steps and avoid falling down or hitting a wall by employing short- range sensors and simple corrective action (such as walking with arms stretched out, and changing direction from one step to another), going at the faster speeds which technology brings into our reach necessitates some longer-range sensors such as lights, or radar. Direction and speed cannot be corrected instantly and must be set in anticipation. The other reason has to do with systemic effects of action we take, effects which have become much more important at a high technological level of living at which the world virtually 'shrinks' and displays many types of systemic interaction and interferences. To view planned action in the light of this systems context again requires the conscious study of longer-range dynamic developments. The time is past, and has not lasted very long, when long- range planning might have been regarded as a fancy means to contribute to individual aims or crude business policies such as 'maximisation of profit'. Today, long-range planning is becoming an integral part of the search for a new cultural basis, new sets of values to be brought into play, and new and flexible responses to a rapidly changing world. It broadens the scope of thought and action in every dimension – not just the time dimension.

Over the past few years, corporate planning has developed considerable sophistication in dealing with the strategic level, i.e. bringing in the concept of a 'decision-agenda' (as Kenneth Boulding calls it) which is to be enriched by as many options as can be developed. The emergence, in the 1960s, of technological forecasting as a means to envisage such options, was crucial in building the fundaments of strategic medium- and long- range planning in corporations active in rapidly evolving sectors of technology. It gave corporate planning its first really long-range perspective, even if only informally. The usual five-year quantitative corporate plans are still to be considered of short-range scope in so far as they usually outline one specific set of operational plans, specifying products to be developed, sales targets, market shares, and so forth. In many technological areas, the time-span necessary to 'freeze' and realise a specific product development is about five to seven years – this would then be the range of operational planning. For strategic planning, technological options may be spelled out roughly within the time-span between laboratory discovery and marketable product – around 15 years, perhaps. But even longer-range time-frames are necessary to grasp societal needs and to plan for ways to contribute to them. The only valid limitation to time-distance in forecasting and planning – the limitation which separates planning from speculation or science fiction – is determined by its impact on action in the present. Planning is long-range thinking affecting action *in the present*.

With increasing sophistication in strategic planning, industry is learning to think not so much in terms of specific technologies, skills, or product lines, but – taking up the challenge of integrative thinking – in terms of the *functions* of technologies and products in a societal (i.e. systemic) context. If, in a major American car manufacturing corporation, a split occurred between people holding that 'we manufacture cars' and others saying 'we move people', the distinction between the function 'transportation' and the product 'car' leads to vastly different behavioural patterns and planning approaches. Thinking in terms of technological functions brings the larger systems – in particular, the 'joint systems' of society and technology, as they may be called – into view and ultimately leads to redefining the corporation's tasks in terms of socio-technological systems design.

This book consists mainly of articles which the author has published ever since his survey of methods and organisation of technological forecasting, undertaken for the Organisation of Economic Co-operation and Development (OECD),[1] has led him to this broad subject. In fact, recognition that it is not so much the sophistication of forecasting approaches themselves which matter most, but the framework of thought and action in which they are employed, has led to a serious concern which finds its expression in this series of articles dealing with many aspects of forecasting, planning, decision making and action. In this search for a meaningful framework to deal with the future, the author has been guided by the basic approaches developed by *Hasan Ozbekhan*[2] and *West Churchman*[3]

[1] Erich Jantsch, *Technological Forecasting in Perspective*, OECD, Paris 1967.
[2] Hasan Ozbekhan, 'Toward a General Theory of Planning', in E. Jantsch (ed.), *Perspectives of Planning*, OECD, Paris 1969; also, 'The Triumph of Technology: "Can" Implies "Ought" ', in S. Anderson (ed.), *Planning for Diversity and Choice*, MIT Press, Cambridge, Mass. 1968.
[3] C. West Churchman, *Challenge to Reason*, McGraw-Hill, New York 1968; also *The Systems Approach*, Delacorte Press, New York 1968.

and also finds himself very much in resonance with the concepts developed by *Sir Geoffrey Vickers*.[4]

For convenience in following the general lines of thought, meandering through various aspects, the chapters of this book (the individual articles) have been grouped under six thematic headings. However, they provide but a loose structure and there was no attempt to discuss them in a comprehensive way.

Chapters 2 to 4 develop a general framework for long-range thinking, its application to the development of technology, and its translation into terms of corporate planning. These chapters thus deal with the basis which also holds for the rest of the book.

Chapters 5 and 6 give a brief survey of some of the principal categories and methodological concepts of technological forecasting, emphasising the current trend toward the development of more systemic (integrative) approaches to forecasting. A more detailed discussion of techniques may be found in the author's OECD survey, already mentioned above.

Chapters 7 and 8 deal with the basic shift from product-oriented to function-oriented thinking which accompanies the introduction of strategic long-range planning in many industrial domains. It leads to major organisational consequences in terms of increased flexibility in the technological response, as well as in the definition of corporate objectives and goals.

These organisational implications come into focus in Chapters 9 to 11. Both the internal organisation of corporations – with emphasis on the principle of *decentralised initiative and centralised synthesis* as well as on flexibility – and the relationship between the corporations and the rest of the societal systems they are part of – the *roles and responsibilities* of corporations – are viewed in the light of long-range thinking. A useful systems approach called *multiechelon co-ordination*, developed by Mesarović and associates,[5] is introduced by way of the author's book review on Mesarović's book.

The subject of roles and responsibilities is pursued further, in more concrete terms, in Chapters 12 to 14. It is ultimately viewed in the light of growth policies in a finite world system, as investigated by Forrester[6] in his approaches to world dynamics, and discussed here in the form of a book review.

Finally, Chapters 15 and 16 attempt to outline a few of the changes that might be introduced to scientific and technological activity in general, and the university, in particular, if science and technology are to be marshalled for a long-range purpose of mankind. These chapters link

[4] Geoffrey Vickers, *Freedom in a Rocking Boat: Changing Values in an Unstable Society*, Allen Lane, The Penguin Press, London 1970; also *Value Systems and Social Process*, Tavistock Publications, London 1968, and Basic Books, New York 1968.

[5] Mihajlo D. Mesarović, D. Macko, and Y. Takahara, *Theory of Hierarchical, Multilevel, Systems*, Academic Press, New York and London 1970.

[6] Jay W. Forrester, *World Dynamics*, Wright-Allen Press, Cambridge, Mass. 1971.

institutional change which is announcing itself in business, to change in some of the other institutions of society which, for better or worse, are the managers of science and technology in our present world.

With this structure of the book, the entire process of rational creative action is spotlighted from forecasting and planning over decision making to institutional and instrumental change, together with the 'links' of relevance, motivation and organisation (*see also Figure 2.3*). Emphasis is placed on the strategic level and to some extent – and as far as thinking has developed there which can be related to the institution of business – also on the policy level. Operational planning with its sophisticated methods (such as CPM, PERT, and other network techniques, or various forms of operations research) has been dropped here – not because it is not important and useful (it certainly is), but because it is already well developed and well known. This book is mainly concerned with the tasks *for*, not the processes of operational planning.

In dealing with its subject in the appropriate broad framework, this book may convey the impression of penetrating far beyond the areas of direct concern to corporate planning. But this lies in the nature of the subject itself: Technology, to paraphrase an old *bon mot*, is too important a matter to be left to the technologists – as planning, too, is becoming too important to be left simply to the planners. Technological innovation as well as planning have become central concerns for top management and, indeed, all management echelons. If corporate policy is conceived here as the regulatory function for the dynamic behaviour of a corporate system, then the latter has to be understood in its relationships with the whole environment. Technological development, in this view, loses its isolation and product-orientation and becomes an aspect of *social creativity*.

Corporations constitute but an intermediate step in social organisation, and neither nations nor even blocs of nations may be considered the highest step. Their policies ought to be conceived and planned as integral parts of a *policy for mankind*, which, in turn, depends on the creative inputs from all subsystems and cannot be decreed from the top, even if such a top would exist at all. The principal aim of this book is to bring institutional roles – and, in particular, corporate roles – in planning for technology into focus. These roles become meaningful only in the broadest possible context of organising science and technology toward a human purpose. The general framework chosen for this book, as well as much of its detailed contents, apply equally to public planning.

With such a human purpose in mind, the relationships between society and technology will be seen in a new perspective. They will no more be considered as linear causal relationships (as the idea of 'technology assessment' still seems to imply), nor as positive, self-exciting feedback loops in which technology engenders more technology, and distortions in the social system – still wrongly perceived as 'side effects' by many – are sought to be remedied by perpetuating the endless technology/counter-technology chain (Dubos). The new perspective will be that of *balance*, and economic and technological change will become aspects of much more broadly conceived political and cultural change. Already, important areas of technological development become increasingly detached from market choice, and subjected to political choice. To achieve balance, will not only require that the powerful momentum of technological growth trends be broken, but will also depend on our response to the new challenge as pronounced by Dennis Gabor: 'Til now, man has been up against Nature; from now on, he will be up against his own nature.'

A FRAMEWORK FOR THOUGHT AND ACTION

2. FORECASTING, PLANNING AND POLICY FORMATION .
3. INTEGRATIVE PLANNING OF TECHNOLOGY
4. TECHNOLOGICAL FORECASTING IN CORPORATE PLANNING

Forecasting is an integral part of the planning process which, in turn, unfolds in the cybernetic process of rational creative action. To engage this process fully, we have to view it at three levels, linked by feedback interaction between them: **policies** (what *ought* we to do?), **strategies** (what *can* we do?), and **operations** or **tactics** (what *will* happen, if we take a specific course of action?) This *vertical integration* of planning and action brings human values and norms into play in an important way.

The *horizontal integration* of planning and action aims at encompassing planning for complex multivariate systems. Technological planning, interacting strongly with nature, society, and man, is viewed as aspect of *integrative planning*, cutting across social, economic, political, technological, psychological, anthropological and other dimensions.

This view of planning and action as *'futures-creative'* (i.e. normative and cybernetic) process – with man assuming the responsibility for regulating large societal, and even global systems – gives corporate long-range planning new facets and profound new significance. Forecasting, in particular technology-oriented forecasting, finds important roles tying in with many of these facets. One of these roles is to provide coupling between the various phases of research, development, and application in the innovation process.

Chapter 2 is based on a presentation in the session 'Approaches to Policy Sciences' of the AAAS Annual Meeting (American Association for the Advancement of Science), Boston, 26–31 December, 1969. It was first published under the title 'From Forecasting and Planning to Policy Sciences' in *Policy Sciences*, Vol. 1, No. 1 (Spring 1970), and is reprinted here by permission of American Elsevier Publishing Company, Inc., New York.

Chapter 3 formed a presentation to the OECD Working Symposium on Long-Range Forecasting and Planning, Bellagio (Italy), 26 October–2 November, 1968. It was first published under the same title in: E. Jantsch (ed.), *Perspectives of Planning*, OECD, Paris 1969, and is reprinted here by permission of the Organisation for Economic Co-operation and Development (OECD), Paris, which holds exclusive copyright.

Chapter 4 is based on presentations to the Technological Forecasting Conference, organised by the Industrial Management Center (Prof. James R. Bright) in Washington, D.C., March 1968, and to the Technological Forecasting Conference, sponsored by the University of Bradford *et al.* in Harrogate (England), July 1968. It was first printed in *Long Range Planning*, Vol. 1, No. 1 (September 1968), and in: G. Wills, D. Ashton, and B. Taylor (eds.), *Technological Forecasting and Corporate Strategy*, Bradford University Press in association with Crosby Lockwood, London 1969, and Elsevier, New York 1969. It is reprinted here by permission of the Editor of *Long Range Planning*.

2. Forecasting, Planning and
Policy Formation

INTRODUCTION

FORECASTING and planning constitute important aspects of the policy sciences to the extent that they deal with the design of policies, which forms part of their content. The advances recently made in their conceptualisation have led to a clearer recognition of the nature of policies and policy making, and of the interaction of forecasting and planning at the policy level with corresponding intellectual pursuits at strategic and operational levels.

Policies are the first expressions and guiding images of normative thinking and action. In other words, they are the spiritual agents of change – change not only in the ways and means by which bureaucracies and technocracies operate, but change in the very institutions and norms which form their homes and castles. 'Policy [consists] in regulating a system over time in such a way as to optimise the realisation of many conflicting relations without wrecking the system in the process', states Geoffrey Vickers,[1] and: 'The object of policy at every level is to preserve and increase the relations we value and to exclude or reduce the relations we hate.' No wonder, then, that political and business bureaucracies favour any degeneration of thinking which permits them to formulate pseudopolicies as emanating from their current operations, from intuitive strategies, and from existing institutions and instrumentalities. Policy analysis is frequently practised today in this self-perpetuating way which does not recognise the logical and practical need for innovative policy design first.

It is simply not enough to apply useful, but fragmented approaches, or to combine them in an arbitrary sequence. Policy making, as the most difficult, innovative, and radical element of the full process of rational creative action, will always deteriorate and vanish from the process first.

[1] Geoffrey Vickers, *Freedom in a Rocking Boat: Changing Values in an Unstable Society*, Allen Lane, The Penguin Press, London 1970.

We deplore the resulting ineffectiveness of our beautiful strategic plans –
which are not becoming implemented – and the impossibility of breaking
out of the linearity of our manifold material and non-material 'vested
interests'.

The present state of the art of policy making, with policy sciences in
their present state not yet inducing decisive change, provides a sad illus-
tration for a statement contained in the 'Bellagio Declaration on Planning':[2] 'Mere modification of policies already proved to be inadequate will
not result in what is right. Science in planning today is too often used to
make situations which are inherently bad, more efficiently bad.'

The only way out of this dilemma is a conscious attempt to bring the
design and implementation of policies into sharp focus. This noble and
urgent task, and nothing else, should form the content of the emerging
policy sciences.

IN SEARCH OF A VIABLE CONCEPTUAL FRAMEWORK: THE PROCESS OF RATIONAL CREATIVE ACTION

Policies, policy making, and policy sciences cannot be defined in an
arbitrary way if they are to have a meaning which ultimately can be
translated into action. The conception and implementation of policies
form an integral and most important part of the *process of rational creative
action*, which has to be properly understood before policies and policy
making can be assigned their roles.

We may conceive of the process of rational creative action as unfolding
in the interaction among four activities: forecasting, planning, decision
making, and action. These activities not only are linked to each other but
are embedded in the succeeding activities so that (*see Figure 2.1*):

Forecasting plus planning = Planning process
Forecasting plus planning plus decision making = Decision making
process
Forecasting plus planning plus decision making plus action = Process
of rational creative action

In other words, forecasting, planning, and decision making have to be
conceived in such a way that they are conformable to human action.

Extending the notions formulated by Ozbekhan[3] for planning, we may
call this approach the *human action model*, as distinct from the mechanistic

[2] The 'Bellagio Declaration on Planning' was formulated by the participants of the
OECD Working Symposium on Long-Range Forecasting and Planning, Bellagio (Italy),
26 October–2 November 1968. The proceedings are published in E. Jantsch (ed.),
Perspectives of Planning, OECD, Paris 1969.

[3] Hasan Ozbekhan, 'Toward a General Theory of Planning', in E. Jantsch (ed.),
Perspectives of Planning, OECD, Paris 1969.

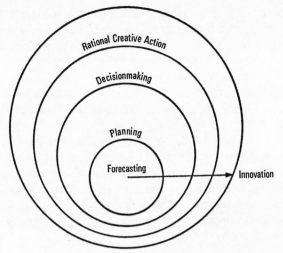

Fig 2.1 The process of rational creative action leading to innovation.

model which corresponds to a deterministic attitude and the type of linear planning and action which govern so many of the current attempts to rationalise action. It may be instructive to compare a few of the general characteristics of the two models, as outlined by Ozbekhan:

Mechanistic Model	*Human Action Model*
Goals given from outside	Selects values, invents objectives, defines goals
Designed to solve specific class of problems	Seeks norms, defines purpose
Internal organisation independent of purpose	Higher order organisation defined by purpose
Controlled by external policy	Self-regulating and self-adaptive
Programmed actions toward given outcome	Regulation of steady-state dynamics through change and governance of metasystem's self-adaptive and self-regulatory tendencies, through policy formation

Policy making is an essential, or even the guiding part of an innovative process based on the deployment of human free will. It can therefore not be part of a mechanistic model – where policies are given by external authorities (e.g., in dictatorially controlled systems). Policy making is *inherently* part of a human-action model. One of the first guidelines for a concept of the policy sciences may thus already be formulated: the ensurance of the human-action-model approach to structuring rationality at the policy level. What this means in detail will become clearer a little later.

HUNT LIBRARY
CARNEGIE-MELLON UNIVERSITY

For reasons of a clear graphic representation, the innovation process from forecasting to action may be stretched out as in *Figure 2.2*. The systemic innovation process appears here in some analogy to the commonly better understood technological innovation process with the phases discovery, creation, substantiation, development, and engineering. In both cases, alternative possibilities are conceived and assessed before a decision is made and a specific way to action is pursued. Of course, there is always some feedback between the successive phases involved – characterised by the arrows of varying thickness in *Figure 2.2*– but a net vector from forecasting to action nevertheless is expected to result.

Recent advances in the conceptualisation of creative forecasting and planning recognise planning as a general instance of human activity and perceive it as an integral part of the decision making process. That forecasting is embedded in planning, has already been stated in early attempts to give it a valid framework.[4]

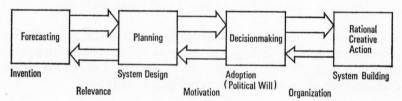

Fig 2.2   The systemic innovation process and key concerns for its phases and the links between them.

Three essential features of the 'new' planning make it radically different from the 'old' (non-creative) planning: (1) The general introduction of *normative thinking* and *valuation* into planning, making it non-deterministic and futures-creative, and placing emphasis on invention through forecasting; (2) The recognition of *system design* as the central subject of planning, making it non-linear (i.e. acting upon structures rather than variables of systems) and simultaneous in its general approach; and, following from the two preceding points; (3) The conception of *three levels* – normative or policy planning (the 'ought'), strategic planning (the 'can'), and tactical or operational planning (the 'will') – in whose interaction the 'new', futures-creative planning unfolds.

The three-level structure not only pertains to planning but serves to elucidate the full process of rational creative action. Ozbekhan's[5] formulation for the purpose of structuring the tasks of planning, may be logically extended to forecasting, decision making, and action:

'There are three levels of functional relations between a plan and the environment:

[4] Erich Jantsch, *Technological Forecasting in Perspective*, OECD, Paris 1967.
[5] Hasan Ozbekhan, *op. cit.* (see footnote 2), p. 153.

(a) Policymaking functions which result in normative planning and are directed toward the search and establishment of new norms that will help define those values which will be more consonant with the problematic environment. In other words, normative planning occurs when the purpose of planning action is to change the value system in order to achieve the required consonance with the environment. The statements of normative planning are derived from values and defined in terms of "oughts".

(b) Goal-setting functions which result in strategic plans wherein various alternative ways of attaining the objectives of the normative plan are reduced to those goals which can be achieved given the range of feasibilities involved and the optimum allocation of available resources.

(c) Administrative functions which lead to operational planning wherein the strategies that will be implemented are ordered in terms of the priorities, schedules, etc., that the situation dictates. Operational planning is that part of the planning structure in terms of which changes in the environment are effected that are purely of a problem-solving nature. (In other words, operational planning need not involve a consideration of value premises.)

The full 'new' planning unfolds in the feedback interaction between these three levels. The 'old' planning reappears here, more or less intact, at the tactical or operational level only. It is no wonder, therefore, that the well-developed procedures at this lowest level tend to distort the full planning process even where its necessity is already acknowledged in a general way.

The above quoted 'Bellagio Declaration on Planning' warns that 'the pursuance of orthodox planning is quite insufficient, in that it seldom does more than touch a system through changes of the variables. Planning must be concerned with the structural design of the system itself and involved in the formation of policy.' And it states as the first rule to be followed that 'the scope of planning must be expanded to encompass the formulation of alternative policies and the examination, analysis and explicit stipulation of the underlying values and norms'.

Figure 2.3 attempts to extend the three-level forecasting/planning scheme (as modified by the author of this book) to the decision making and action phases in order to gain an integral view of the structured rationalisation of creative action. The normative character of the process of rational creative action is indicated by the dominance of norms, which are derived from values. This is equivalent to stating that the depicted scheme recognises policy making as the guiding function of the entire action process.

The full scheme integrates (in this stretched-out representation of a configuration which, in reality, is to be viewed as in *Figure 2.1*) four vertical columns and three horizontal levels. The four vertical volumns represent the four activities, already recognised in *Figures 2.1* and *2.2*, but structured in three levels:

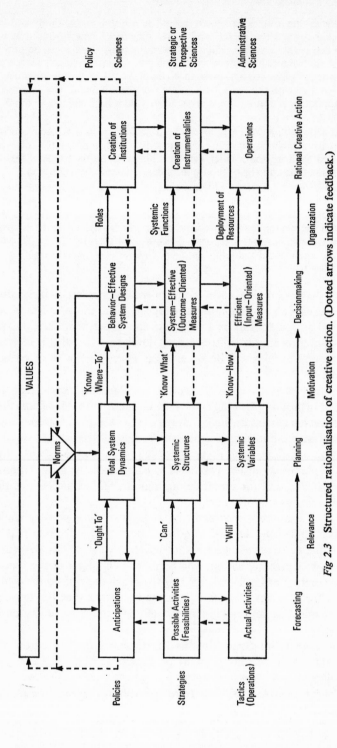

Fig 2.3 Structured rationalisation of creative action. (Dotted arrows indicate feedback.)

—*Forecasting* deals with the invention of anticipations[6] and of possible (feasible) activities fitting them, and with the probabilistic assessment of (assumed or real) actual activities. Interaction among the three levels is by continuous feedback processes (or, at least, multiple iteration), but the downward pointing full arrows are supposed to emphasise the basic normative character of forecasting; for example, a compilation of possible activities cannot determine anticipations, but anticipations, valuated as 'good', determine the spectrum of possible activities to be considered, within the range of feasibility.

Forecasting and planning are tied together particularly closely (since creative planning is based on invention through forecasting), and they shape their policy and strategy constructs together in feedback or iterative processes with no really dominating direction. These ties may be called *relevance*, expressed by the same 'ought to', 'can', and 'will', that also characterise Ozbekhan's view of the planning levels.

—*Planning* deals with system design at the levels of total system dynamics, system analysis to define effective changes in system structures (goals), and changes of variables in given system structures. The feedback interaction among the three levels is of the same nature as with forecasting. Planning provides the information basis, in dynamic terms, for decision making.

In a normative framework of rational creative action, planning has to provide for links to decision making which may be called *motivation*, expressed by the information values of 'know-where-to', 'know-what', and 'know-how'. The current dominance of 'know-how' in the motivation of decisions (e.g., for technology) illustrates the grossly distorted balance among the three levels of planning and decision making. But information on the 'know-what' may also be meaningless, if it is not scrutinised in the light of criteria provided by policies and making it possible to distinguish between 'good' and 'bad' strategies.

—*Decision making* deals with the recognition, again in a normative way, of system designs which can be expected to be effective in changing the dynamics of the system in the direction of preferred anticipations and increased dynamic stability, of system-effective measures (whose outcomes do not upset systems in the same way as, for example, the introduction of technology for specific 'single-track' purposes frequently upsets the systems of human living), and of efficient measures to do well what has been decided to do. The predominance of efficiency and decision making structures focussing on it, bears testimony to the current distortion in the decision making process.

Decision making leads to action by means of *organisation*, which focusses at the three levels on the definition of roles (within an adopted dynamic system design), on systemic functions (e.g., the functions of technology in social systems – communication technology, information technology, health technology, etc.) – and on the deployment of resources of a material and non-material nature (including the available basis of science and technology, human skills, etc.)

—*Rational creative action*, finally, deals with the creation of institutions,

---

[6] Anticipations are 'intellectively constructed models of possible futures' (H. Ozbekhan); Bertrand de Jouvenel calls them 'futuribles', Herman Kahn 'alternative world futures'.

the creation of their corresponding instrumentalities,[7] and operations within and through these instrumentalities. The growing recognition that institutional change has to precede all other change in social systems, is an expression of frustration by forecasters, planners, and decision makers alike who see no way to implement their ideas and preferences in the rigidified structures of existing institutions and instrumentalities. It is also an expression of the a-priori necessity of policy design human creative action.

These four vertical columns, the activities embedded in the process of rational creative action, do not constitute disciplines, or even inter-disciplines in the conventional sense, but – as has already been stated above – general instances of human activity. They make use of various disciplines and inter-disciplines, such as logic, psychology, political science, system theory, organisation theory, information technology, the entire spectrum of behavioural sciences, and many more. These disciplines and inter-disciplines constitute *aspects* of these four activities. More, and different, aspects come into play if we look at the horizontal layers as defined by the three levels of policies, strategies, and tactics (or operations):

—At the *policy level*, we gain a first view of a viable conceptual framework for the policy sciences. We may say, the *policy sciences deal with the horizontal unfolding, as general instances of human activity, of forecasting, planning, decision making and action at the policy level.*[8] What is meant thereby, is more than strengthening the horizontal coupling between the four activities. It is their conformity in conception and realisation, even their identification, which must be aimed at. The invention of anticipations, the design of social systems with preferred dynamic characteristics, the definition of roles for and the creation of institutions, and the feedback interactions between them, have to be seen as one single task which we may also call *ensuring the human action model in structuring rationality at the policy level*. The disciplines and inter-disciplines of which the policy sciences constitute specific aspects, include philosophy, psychology and social psychology, sociology, law, political economy, anthropology, system sciences, and others. The emphasis is on *design*: system design, policy design, design of institutions; in other words, on intellective innovation.

[7] The terms 'institutions' and 'instrumentalities' are used here in the meaning developed in a seminar on long-range planning which H. Ozbekhan conducted together with the author (European Forum, Alpbach, Austria, 23 August–8 September 1969): 'Education' is an institution, schools and universities are instrumentalities; 'family' is an institution, 'marriage' an instrumentality; 'government' and 'religion' are institutions, specific forms of government (e.g. parliamentary democracy, or constitutional monarchy) and the various churches are instrumentalities; etc. This notion of institution corresponds to Talcott Parsons' view of institutions as 'crystallisations' of values in society; however, institutions are viewed as taking active part in a cybernetic process here, as becomes clear from *Figure 2.3*.

[8] The human activities of policy making and policy implementation are, of course, not themselves part of any science. Policy sciences refer to the logical, methodological, organisational, etc., aspects, i.e. to the theoretical framework.

—At the *strategic level*, we find a conceptual framework for what we may tentatively, and for lack of an accepted term, call the *strategic* or *prospective sciences*, which have not yet become recognised as such. However, the task of dealing integrally with the conception of possible activities, their implications for system structures and system-effectiveness, and the creation of instrumentalities – in other words, the human action model of structuring rationality at the strategic level – has already come into better focus than the corresponding task at the policy level. This is particularly true for advanced-thinking industry, mainly in the United States, and for the development of military and space technology. The disciplines and inter-disciplines of which aspects come into light at this level, include sociology, technology, system sciences (in particular system simulation), decision theory, and the entire spectrum of socioeconomic or, more general, integrative[9] approaches (comprising, *inter alia*, Planning-Programming-Budgeting, a possible generalised 'social cost-effectiveness' approach, and Bertrand de Jouvenel's SPES – 'Stratégie Prospective de l'Economie Sociale').[10] The emphasis is on *analysis:* system analysis, policy analysis (which does not belong to the policy-forming level!), need analysis, institutional analysis, market analysis. The results of rational creative action at the strategic level express themselves mainly in social and technological innovation.

—At the *tactical* (*operational*) *level*, the corresponding task of integrating probabilistic forecasting related to actual activities, the changing of system variables in given system structures, and efficient ('economic') operations is pursued in a most effective way by the *management*[11] and *administrative sciences*. Their concepts dominate the entire process of rational (but, in this case, non-creative) action to such an extent that linearity and sequentiality prevail, and that institutions and even instrumentalities (the latter particularly in the public domain) are perpetuated by this process. Disciplines and inter-disciplines providing aspects for this level, include, above all, 'correct economics' (with linear economic forecasting, econometrics, Keynesian growth economics, cost-benefit assessment, and discounted cash flow approaches), administration, operations research, resource allocation and deployment, and others. The emphasis is on *expansion:* expansion of actual activities, of system variables, of resources, markets and operations in general. The result is conservation of systemic structures and increasing 'systemic clogging'.

It is obvious that the unfolding of rational creative action is weakest at the policy level, in the legitimate realm of the policy sciences. Nevertheless, the full process of structuring the rationality of creative action cannot

[9] The term 'integrative' has come to denote approaches to forecasting, planning, and decision making, which cut across many dimensions, in particular social, economic, political, technological, psychological, and anthropological dimensions.

[10] Bertrand de Jouvenel, *Arcadie, essais sur le mieux vivre*, in the 'Futuribles' series, SEDEIS, Paris 1968.

[11] More imaginative frameworks, recently developed for the management sciences, include to an increasing extent notions belonging to the strategic level. It is conceivable and even desirable that the management sciences will lose their present image of dealing primarily with linear change and operations. In this case, 'management' may ultimately acquire notions coming close to the management of rational creative action in general, i.e. belonging to all three levels.

function properly, cannot even start, as it ought to, from values and norms, as long as the top level – the policy level – of human action is partly or fully neglected.

The correct logical order of the process of rational action according to *Figure 2.3* is to proceed from the left to the right, and from the top down. No system-effective measures can be decided without at least tentative knowledge about a behaviour-effective system design. No instrumentalities can be logically created without having first created institutions. Today, however, we are becoming increasingly aware of our proceeding in the wrong direction, especially the direction from the bottom up. We may call this 'system-immanent' action, the result of which is the rigidification of existing system structures, and short-circuiting of normative thinking.

Figure 2.4 presents three characteristic types of *degeneration* of the process of rational creative action: the *'bureaucratic' type* (seeking conservation by focussing on merely operational action, and influencing norms accordingly);[12] the *'technocratic' type* (pursuing the pseudochallenge of technological and instrumental promiscuity, and influencing norms accordingly);[13] and the *'futurocratic' type* (emphasising the need for systemic change and simultaneously proving its impossibility by following a rationale derived from the operation of given system structures). All three examples represent essentially non-normative action, because the norms are strongly influenced by the action itself, and less by values.

The tentative concepts of the policy sciences, as they have emerged so far, in many cases tend to enhance the degeneration of the process of rational creative action. The principal reason for this, as paradoxical as it may sound, is a certain neglect of policy making in its proper meaning of policy *design*, and of action at the policy level in general. Another reason is a babylonic confusion among issues at all three levels – policies, strategies, and operations – and a fragmented or ill-structured approach to all ingredients of rational political action.

If the concept of the policy sciences, sketched in this chapter, is accepted, a number of far-reaching implications for planning, organising, and implementing rational creative action have to be taken into account. A few of them will be briefly explored in the following section.

TASKS FOR THE POLICY SCIENCES

From the foregoing it becomes clear that the principal tasks for the policy sciences will be found in the development of concepts and measures to

[12] The German word 'Sachzwänge' (compulsion through circumstances) is the precise expression of such an attitude.
[13] The 'rule of the technostructure', before which John K. Galbraith warns in his book *The New Industrial State* (Boston 1967), would result from such a setting of goals in the absence of the full normative thinking and planning process.

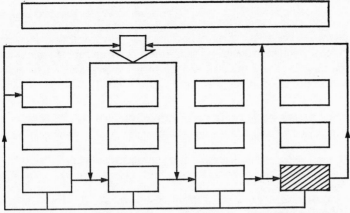

(a) 'Bureaucratic' Type. Action here is neither rational nor creative. The result is the rigidification of current structures, activities, and modes of operation.

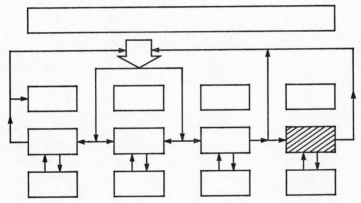

(b) 'Technocratic' Type. The result is the indiscriminate (non-normative) creation of new structures and instrumentalities: 'Can' implies 'ought,' as evidenced by 'the triumph of technology' (Ozbekhan).

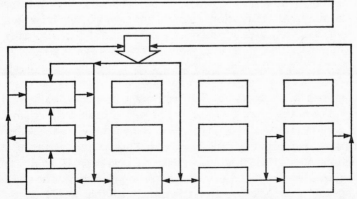

(c) 'Futurocratic' Type. The result is the futile attempt of 'system-immanent' change through measures at the operational, or at most, strategic level only.

Fig 2.4 Types of degeneration of the process of rational creative action.

strengthen the *links* (the arrows in *Figure 2.3*) at the policy level and to make them effective in ensuring conformity between forecasting, planning, decision making, and action, and between values and these activities. These links are the following ones:

(a) Norms, linking values to the process of rational creative action.

(b) Links among the four activities of the process of rational creative action, at the policy level:
—Relevance between forecasting and planning (the 'ought to')
—Motivation, leading from the planning process to decision making (the 'know-where-to')
—Organisation, leading from the decision making process to action (the definition of roles)

(c) The feedback loop between policy formation and action, dynamic values and norm configurations, and induced changes in the environment.

Norms bring values into the play in an explicit way. They are rules for behaviour, derived from values, guiding policy formation and action. If policy making is viewed, as in the preceding chapter, as the process of searching for those norms which will help define value configurations enabling us to deal with problematic situations, its essence is the simulation of the consequences which will follow from the adoption of alternative sets of norms. Ethics may be an aspect of norm-seeking, but it is traditionally a more or less static concept. The approach of the policy sciences to norms is new in that it is dynamic and active, and that it is guided by the state of the environment and by the dynamics of environmental systems – in other words, by acts of human free will to shape the environment. To link norms to values on the one hand, and to the environment on the other in an explicit way, and to study their effect on system designs and system dynamics, is a new task to be explored and developed by the policy sciences.

Relevance between forecasting and planning at the policy level is provided by normative statements, by the 'ought to'. Anticipations become causative anticipations (Ozbekhan) through this relevance link. The link, with which policy sciences have to deal here, is actually very complex in its nature. It involves the modelling of a dynamic situation, the invention of new system designs and their modelling, dynamic simulations of new system designs and the attempt to match future system states with anticipations (data bases for intellectively constructed models of the future). For most practical uses, this will be a sizeable task, in particular when a certain degree of quantification is aimed at. Feedback between forecasting and planning will be very intense in this phase of policy formation, and anticipations may have to change considerably in the process, involving new feedback with norm configurations. The task is not to translate hedonistic constructs into action – this would be a futile attempt – but to change, through new designs, the dynamic behaviour of complex social systems so that better consonance with desirable future system states (anticipations) and norm configurations may be achieved. This is an extremely difficult task for the policy sciences, one for which methodological developments are still in their infant stage.

Motivation carries the planning process over to decision making. Two ingredients are essential for this: The projected outcomes of the adoption of plans must be believed by the decision makers; and these outcomes must be in consonance with the norms followed by the decision makers or the part of society they represent. Again, this is a most difficult task. Man, perhaps tied to his biological inheritance, is inclined to react to changes in the environment, not to act in a rational, creative way. In other words, he tends not to believe anticipated outcomes to the extent that he would take action in an anticipatory way. This will be even more true, the more counter-intuitive such anticipated outcomes become, as may indeed be expected for the types of complex social systems we are dealing with.[14] Nevertheless, motivation is becoming ever more urgent, since the time constants in the system dynamics are becoming dangerously small. Where, during the crossing of 'thresholds' (J. Huxley) in earlier phases of mankind's psycho-social evolution, reaction to slowly changing environmental situations was possible, it may now lead to dangerous excursions. As a matter of fact, the time left to us for anticipatory action in a number of critical areas which might have some effect in alleviating (not preventing) crises, is becoming very short – in many areas it is between 3 and 15 years only.[15] Even if plans can be worked out in time, motivation may not be achieved in the short time-spans left. One of the greatest potential contributions of the policy sciences could be made by exploring and designing means to accelerate motivation and make it more effective. Conceivably, quite new communication processes will have to be devised in problem areas affecting all or large parts of society. At the same time, the second ingredient of motivation, consonance of norms, will put the traditional democratic processes, in particular participative democracy, in entirely new perspectives. The key issue will probably be how to motivate and accelerate a widespread *'prise de conscience'* in an explicit way. (At present, such a *'prise de conscience'* is taking place among part of society, in particular among younger people, in an implicit or ill-focussed way only; also, the process is apparently too slow.) Policy sciences will find a most important task in designing mechanisms for translating the 'know-where-to' resulting from planning – at first by clarifying where some societal or global systems are currently drifting to, or are being pushed to – into a massive and overwhelming *'prise de conscience'* which would lead to

[14] Jay W. Forrester, 'Planning Under the Dynamic Influences of Complex Social Systems', in E. Jantsch (ed.), *Perspectives of Planning*, OECD, Paris 1969.

[15] Hasan Ozbekhan, 'On Some of the Fundamental Problems of Planning', *Technological Forecasting*, Vol. 1, No. 4 (1970).

decisions on a scale and of a nature politicians and business leaders are unable to imagine today.

Organisation, leading from the decision making process to action, forms the last of these crucial links. The type of action it ought to lead to is the creation of new institutions or, to some extent, perhaps the changing of existing institutions. This is action of the rarest and most difficult kind, action which we usually try to get around. Only with explicit and believed policies in hand can we attempt this step – and only when policies and policy design can be viewed in sharp focus will the inevitability of action at the policy level become clear and compulsive.

Organisation at the policy level means the definition of roles (of institutions and their instrumentalities), projected against a changing environment. People are normally part of existing institutions and instrumentalities, and their mandate to plan and act on behalf of social systems, stems from such existing structures. How can these people then change their own institutions and instrumentalities? There are two approaches possible and both are tried out in a number of cases today, especially in American industry: gaining flexibility by acting from the outside or from the inside of existing structures.[16]

The second approach is more effective, although more difficult. It requires an extraordinary intellectual effort to 'go outside' existing structures in an abstract way, by going to the next higher level of abstraction. This level is the level of values, and the links to intellectual innovation are the norms. It is an entirely new type of politician, manager, and university administrator who would be able to see his principal task in thinking at the level of values and thereby to order his views in such a way that he is capable of reflecting on the role of his organisation in the light of anticipated changes in the environment (which he may influence himself, or not).

Finally, the *feedback loop* between policy formation, the creation of institutions (and their instrumentalities), the changes thereby induced in the environment, and the dynamics of values and norm configurations – an almost totally unexplored field – must become a major theme for the policy sciences. We do not know to which extent we influence our own values by shaping our environment. It is a fact that values do change in reaction to changes in the environment. (An expression of this is that institutional change normally occurs in times of major crises or catastrophes.) It would be decisive to know in what way values change, or may be brought into play, in response to anticipated changes in the environment. With the fast pace of system dynamics in the world today, such anticipatory changes in our values or value and norm configurations may hold the only hope for planning and implementing action in time, i.e. when a marked effect of such action can still be expected.

[16] See Chapters 9 and 10 in this volume.

A few more general observations may be added to throw some light on the implications of adopting a framework for policy planning and action:

Policies are normative expressions of future states of dynamic systems. Therefore policies can only be formed in a meaningful way if the boundaries and structures of such systems are clearly perceived. The roles of institutions can only be reflected upon in the context of the systems to which the institutions and their instrumentalities belong. This leads to vertical and horizontal integration of thinking and action going far beyond the fragmented institutional roles recognised today.

Vertical integration occurs because rational choice is only possible from a viewpoint at the next higher level of abstraction. If we take the challenge of rational creative action seriously, we cannot stop at a lower level than the level of society at large.

Horizontal integration is necessary because we are dealing with total system dynamics, not with the optimisation of subsystems (which would ultimately defeat the ends we envisage for the total system). For many problems, national, regional, or even world-wide systems form the framework for action in this sense. This implies, for example, that we have to think of new institutions and instrumentalities going beyond the artificial systems defined today by nation-states, and have to try to become capable of focussing on critical problems at a very large, or even world scale.

The notions of fragmentation and suboptimisation are incorporated in the crude classical notion of 'free enterprise', in the dominant idea of the university interacting passively with society, and in the interpretation of 'academic freedom' as the way in which the university of today is the mercenary of government and industry.

I have discussed elsewhere[17-19] my views that both industry and the university have to become political institutions in the broadest meaning, i.e. active participants in the planning for society. In particular, science and technology can be brought to interact with society in a truly normative way only if all three institutions in the triangle: government–industry–university change considerably (or are replaced by new institutions).

It is quite obvious that new notions will emerge in this process for the institutions of government, economy, education, etc. I see industry developing an active role in inventing, planning, designing, building, and perhaps operating large-scale socio-technological systems. I see universities

[17] Erich Jantsch, 'Technological Forecasting for Planning and Institutional Implications', in *Proceedings of the Symposium on 'National R & D for the 70's'*, National Security Industrial Association, Washington, D.C. 1967.

[18] See Chapter 10 of this volume.

[19] Erich Jantsch, *Planning for the 'Joint Systems' of Society and Technology – the Emerging Role of the University*, Massachusetts Institute of Technology, Cambridge, Massachusetts, May 1969. Extracts from this report have been published under the same title in *Ekistics*, November 1969, and a short version under the title 'To Help Students More Help Society More' appeared in *Innovation*, No. 4, August 1969. See also Chapter 16 of the volume.

participating in the competition with socio-technological system ideas and designs, and becoming the central agent for ensuring the self-renewal of society (in the profound meaning given to this notion by John Gardner). And I see government assuming a role of organising and synthesising a decentralised inflow of creative ideas and plans rather than administrating, in a 'closed' loop, existing structures and institutions. Government will also have to be instrumental in developing new forms of democratic inter-action; the latter will certainly become one of the major themes for the policy sciences.

If Daniel Bell[20] is right in characterising the post-industrial society, toward which we are headed, by the emergence of intellectual technology and the dominance of knowledge institutions – and I think he is right – we have to design systems, and roles within systems, in such a way that we can bring this intellectual technology to benefit society in the best way. The key notion is flexibility – flexibility in ideas, plans, decision making, and, above all, flexibility in institutional and organisational forms which have to be devised in such a way that they will facilitate, not hinder, continuous self-renewal.

To bring intellectual technology into play in an effective, that is to say in a rational and creative, way, and to put it into a framework of thinking and action capable of a high degree of flexibility, is in my view the central subject of the emerging policy sciences.

[20] Daniel Bell, 'The Balance of Knowledge and Power', *Technology Review*, June 1969.

3. Integrative Planning of Technology

TECHNOLOGY, from the mastering of fire and from the earliest beginnings of primitive weapons technology, is the instrument which enabled man to depart from purely biological evolution and to enter a psycho-social phase of evolution which he shares with no other creature of Nature. In its origins and use throughout the millennia, technology was the tool that permitted man to emancipate himself from Nature, to domesticate Nature – in brief, technology became the tool for Nature engineering, as we might call it today. But it also became instrumental in the attainment of higher stages of psycho-social evolution up to the complex and integrated society of today.

We may thus, in a simplified picture, distinguish between four basically different areas amenable to engineering – areas to which we may apply our capabilities of planning, design and operation through which we can actively introduce change: Nature, man, society, and technology (*see Figure 3.1*). It is of the utmost significance that technology appears here not only as change agent, but in its full ambivalence as an instrument for effecting change and an element of planetary development in its own right. Technology is today the predominant means to engineer Nature, society, and (to an extent which is only marginally realised at present) man himself; but technology is also the possibility to build an artificial world supplementing or even replacing Nature and introducing any number of man-made elements into the human environment. It is this ambivalence of technology which forces us today to attempt control of the development and application of technology in an integrative way, taking into account the full scale of inter-relationships of technology engineering with the other forms of engineering – Nature, human and social engineering – with which it forms an indivisible system.

The decisive factor in the current development, if we still consider the simplified picture given in *Figure 3.1*, is the rapid and almost independent growth of the 'Technology Engineering' area *per se*, and its one-sided influence on the other engineering areas, primarily on the slowly developing area of social engineering and on the area of Nature engineering which is frequently disregarded in planning. Technology engineering is affecting the insufficiently known area of human engineering in an unknown, but certainly important way. The uncontrolled growth of this one element in our overall system, technology, is now about to assume the characteristics of cancerous growth, disturbing and repressing the healthy development of the other parts of the system.

Another important factor is the powerful *positive feedback* which technology, in its role as change agent, obviously introduces into Nature and social engineering, and also into human engineering, if we consider the population explosion as a result of health and nutrition technology.

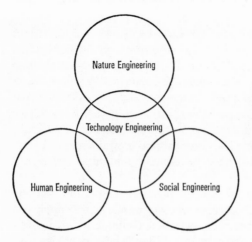

Fig 3.1 The interaction of four different forms of engineering. Their 'autonomous' development endangers the stability of the total system.

The engineering possibilities in these areas are multiplied, the objectives become diversified, changes once introduced become irreversible and give rise to more changes, the dynamics of change are accelerated – in brief, the system (which is equivalent to man and his environment) tends to develop dynamic instabilities which create the danger of catastrophical excursions and subsequent stabilisation on a much lower level of psycho-social evolution only, characterised, *inter alia*, by a lower state of integration.

There can be no doubt that this positive feedback, provided by technology, has contributed enormously to the acceleration of mankind's development over the past millennia and particularly over the past few centuries, and has introduced the vast variety of possibilities which enables us today to choose among a wide spectrum of anticipations ('possible futures') and to formulate our goals in terms of 'quality of life'. However, there are numerous indications pointing to the fact that the destabilising effect of positive technological feedback, for long restricted by the rate of technological innovation and by the available resources, has already seriously distorted development on our planet and is creating the danger of complete loss of control. We notice the alarm signals in the form of various *alienations:* the alienation of man from technology, from society and also from Nature, the alienation of society from technology, etc. Herbert Marcuse's accusation of technology (to quote only one example of some actuality) as the non-political rationalisation of totalitarianism refers to such an alienation caused by a belief in the autonomous development of technology, which is bound to dominate the Nature–man–society–technology system.

What negative feedback can be introduced to counteract the powerful positive feedback from technology and stabilise the dynamic development on our planet sufficiently to avoid dangerous fluctuations? Nature can play such a role only locally and marginally. It could again become a major restrictive factor only after the population explosion has led to a catastrophical situation (for example famine reducing the world population, etc.). The real and undivided responsibility for the continuity of mankind's evolution falls back on man himself and on the amplification of his capabilities through the institutions of society. There can be no doubt that it is in mankind's power to obtain the necessary control through the implementation of integrative planning. The emergence of a new planning philosophy, emphasising the normative aspect of planning and the possibility of actively shaping the future on our planet, has not come too early to check a development which is already characterised by the unmistakable signs announcing a major crisis.

THE 'BI-POLAR' SUBSYSTEMS AND FUNCTIONS FOR TECHNOLOGY

A closer look at the Nature–man–society–technology system shows that it can be broken up into six 'bi-polar' subsystems, each of which represents the integration of two of the four basic elements (*Figure 3.2*). Three of these subsystems contain technology as one of their elements: these are the Nature-technology, the man-technology, and the society-technology subsystems. Three more leave technology out: these are the Nature-man, the man-society, and the Nature-society subsystems.

Control over a specific systems component can be achieved only if we go to the next higher level of abstraction and formulate our objectives at that higher level. We can satisfy this generally valid rule, particularly suited to our purposes, by looking at the *outcomes of technology* within the above 'bi-polar' subsystems. In other words, we look at the *functions* technology performs in these subsystems and we become detached from technology in two important ways: (1) we are now free to consider different technologies contributing to these functions, and to compare the merits of these contributions – and in turn the merits of specific technologies in the

context of such a 'bi-polar' subsystem; and (2) we can now apply normative thinking to functions of technology (needs, impacts, side-effects,
etc.) in sufficient transparency to bring our human value systems into the
play.

One decisive factor of such a function-oriented strategic planning
framework, as we shall now call it, is the presentation of alternatives based
on different technologies; strategic decision making on this basis then
consciously chooses between a variety of feasible technologies, selecting
some, and discarding the others. The criteria for decision making, based
on the comparison of characteristics which the strategic alternatives have
in common, have now shifted from relatively superficial and peripheral
features of specific technological developments, such as material and non-
material resources and economic and performance estimates – because
technologies themselves cannot be compared with each other – to the

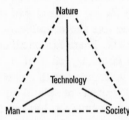

Fig 3.2 The Nature-man-society-
technology system can be broken
up into six 'bi-polar' sub-systems
which facilitate the definitition of
systemic functions. Three of
these sub-systems involve technology (full lines), three do not
(dotted lines).

outcomes of different technologies in a systems context, which, of course,
includes economic and performance aspects. These outcomes, it should be
noted, can be readily compared with each other in qualitative terms, with
the possibility of developing quantitative frameworks for comparison. A
straight comparison between the traditional internal combustion engine
car technology, and the electric, hybrid, steam-engine, etc., car technologies
would emphasise such characteristics as unit cost, operating costs, and
various performance parameters (speed, acceleration, etc.). Consideration
under the function 'urban transportation', in contrast, would not neglect
the above characteristics, but attempt to bring out clearly the effects on the
society-technology system (air pollution, noise, etc.) and on the Nature-
technology system (degrading of fossil fuels through combustion versus
stored electricity from primary nuclear energy, etc.). A wider range of
technological alternatives would perhaps disfavour the car as a possible
urban transport technology altogether. The choice made is then, above all,
that for a specific quality of life.

We have thus obtained both a higher degree of comparability at
strategic level and the possibility to plan in terms of the quality of life.
Another important factor is that we are planning not only for the development of a specific technology, technological product or process, but for the
integrated chain of technological development and its effect in the system –

or, to put it in Harvey Brooks' terms, for both vertical *and* horizontal technology transfer.[1] With the present shift of emphasis from vertical towards horizontal technology transfer ('making better use of existing technology') this integrated planning will become ever more important and inevitable. We may even go considerably further and view technological forward planning now as the *integrated planning of change through technology* in one of the 'bi-polar' subsystems, for example comprising the planning for social change and for new institutions in the society-technology subsystem. It is evident that this new look at technological planning entails tremendous consequences.

A further important feature of function-oriented strategic planning becomes clear from the above: this is the enforcement of *long-range forecasting and planning* in accordance with the long-range objectives that are implied in the functions for technology. Not the emergence or planned realisation of new technology determines the time-frame (as it would, if we focussed our attention solely on technology itself), but the change introduced by technology into the system. It is not surprising from this viewpoint that 50 years or more is to be considered the proper time-frame for strategic technological planning in many contexts. Particularly long time-frames can be readily established for the Nature-technology subsystems and are already, implicitly or explicitly, governing such developments as nuclear power (e.g. emphasis on fast breeders to replace the low-burning reactor generation).

A closer look at the three 'bi-polar' subsystems comprising technology may give us a more precise idea of such a function-oriented framework of thinking:

(1) *The Nature-technology subsystem:* The functions in this subsystem are at the historical origins of technology. In man's fight against an inimical environment and against scarcity, he used technology first with striking success. We might attempt to formulate a few functions in this subsystem in the following way:
    (a) *'Domestication' of Nature* for man's purposes: Agriculture has provided the outstanding success in this area for the past 5,000 years, including the more recent modification from a closed plant/animal system with dirt fertilisation, to an 'open' system receiving inputs in form of chemical fertilisation. It is significant that the necessary modifications which ought to be introduced today to

---

[1] Vertical technology transfer signifies the movement from a scientific principle over elementary new technology to new technological products, processes, systems, etc.; horizontal technology transfer signifies the diffusion and application at a given level of the vertical transfer process, in particular the diffusion of technology (e.g. computer technology) and its application to new ends (e.g. specific industrial control and automation purposes) as well as the creation of new services (e.g. information retrieval services).

adapt agriculture to tropical climates are not pursued with the same vigour and the same success. New forms of biological food production, such as the growth of Single Cell Protein on petroleum or other organic substrates, may soon have to be integrated with the agricultural system. Ocean harvesting (which does not yet constitute 'domestication') could be extended to controlled mariculture, with greatly increased specific outputs. Hydro-power, finally, has to be considered as a type of 'domestication' which was only partly successful, in so far as it affected the natural and ecological stability of certain regions; also, it reaches the limits of exploitation relatively soon.

(b) *Ecological stabilisation,* or *the conservation of the reproductive capabilities of Nature:* In respect of this function (which is only now becoming recognised), the positive feedback of technology has already had some devastating effects. A number of irreversible changes have been introduced – some of which have been going on for centuries – and led to the destruction of forests and fertile lands, the eradication of animal species, drastic change in regional animal ecology, the pollution of lakes and the (partly irreversible) destruction of their natural biological purification mechanism, the considerable increase in the carbon dioxide contents of the air over the past few decades of world-wide industrialisation, the creation of drought zones due to high local energy consumption, the lowering of the ground water level due to over-exploitation and the resulting increase of salinity (in coastal regions), etc. Many of these irreversible, or only slowly reversible changes in Nature, have introduced dynamic processes widening the gap from the original stable state much further – a most undesirable and unforeseen positive feedback, that may eventually leave us only the possibility of supplementing a partly destroyed Nature by an artificial world, and thus turning us back to technology which was at the origin of the destabilisation.

(c) *Utilisation and availability of natural elements, compounds and materials:* It is obvious that here natural reproductivity cannot cope with the rate of utilisation by man. The emphasis ought, therefore, to be placed on the exploitation of new natural resources (bottom of the ocean, etc.), recovery cycles, e.g. for metals, possibilities to supplement and replace scarce natural materials (as has been done with some building materials), and on the proper use of natural compounds and materials according to their most valuable potential (upgrading of the macromolecules available in petroleum for purposes of petrochemistry and food production, etc., instead of downgrading them for combustion purposes).

(d) *The physical adaptation of the natural environment:* The possibilities of water desalination, air conditioning, transport systems, and new forms of shelter, open up a number of possibilities to make new parts of our planet habitable. It seems, however, that with this there is also an irresistible temptation at work to over-shoot the target and plan for domination by a more or less completely artificial environment which technology engineering lets appear to be feasible in the near future. Living in R. Buckminster Fuller's geodomes (conceived so as to float in the air even) in eternal sub-tropical climate, with Commander Jacques Cousteau under the surface of the sea, with some Japanese city planners on the surface of the sea, or with Fritz Zwicky and others on the moon and on planets, certainly would, while creating physically habitable environments, on the other hand interfere drastically with the society-technology and the man-technology subsystems, probably over-stepping the absolute limits set by them.

(e) *The creation of a 'biological landscape'* as our final example of possible functions within the Nature-technology subsystem, is perhaps the most complex task technology faces, it is also closely related to aspects of the Nature-man and Nature-society sub-systems. The target would be, in short, to replace the ugliness of the quickly spreading urban-industrial landscape by the equivalent of the landscape which, in rural societies, is determined by agri-culture. Although agriculture, as a form of Nature engineering, has changed the original natural landscape considerably, it did not break man's relationship with Nature. On the contrary, looking over fields and pastures, we have a unique feeling of being at peace with Nature, of having shaped an 'anthropomorphous' Nature. In contrast to that, the artificiality of our urban-industrial landscape and the visible predominance of pure technology engineering, do not strike us in the same way, but give rise to a feeling of alienation.

(2) *The society-technology subsystem:* This subsystem is now quickly coming into the focus of attention of governments, of industry and of the academic community. Technology is accepted today as the most powerful change agent at work in our society, as regards both its developed and its developing part. It is in this subsystem that functions are most readily recognisable and recognised. Suffice it to name a few of them in arbitrary order, without going into more detailed discussion: communication, transportation, education, energy generation and transmission, public health, automation, security, economic develop-ment, population control, food production, urban development and rehabilitation. It is evident that many of these functions overlap or are mutually dependent on each other, such as the functional complex communication – transportation – urbanisation – automation (im-proved means of communication and automation, such as the home

computer console and satellite educational TV, may render traditional transportation needs obsolete, and influence urbanisation accordingly). The effectiveness of global population control evidently influences greatly the requirements for many other functions. In the society-technology subsystem, the functions represent a less rigid planning framework than those in other subsystems. This is due to the continuous rapid change of society itself, i.e. to the pace of social evolution. The above-mentioned example already gives a hint at the 'mobility' of the transportation function where the focus may shift, in a few decades' time, from present transportation needs for professional purposes to recreational and similar purposes. Or, with the increasing automation, the communication function may take on new aspects such as exercising one's profession by communicating from home with professional centres.

An important aspect of technological development is its conscious planning and utilisation for the purposes of improved social engineering (information and feedback systems such as the 'daily voting' system, mass media, urban experiments, etc.).

The United States Government's Planning-Programming-Budgeting System (PPBS) and a variety of industrial planning schemes and 'innovation emphasis structures' have already made an impressive start in function-oriented strategic planning in the framework of the society-technology subsystem.

(3) *The man-technology subsystem:* Frequently neglected, or subsumed under the society-technology subsystem, this subsystem is likely to provide the most stringent negative feedback elements to control technological development in an integratively planned overall system. The relationship of man to technology is different from society's relationship to technology. The one-sided concern over technology's functions *vis-à-vis* society have led to the restriction of man's freedom (increased urban traffic leading to reduced personal mobility, etc.), to the invasion of his privacy (street noise, radio and television, etc.), to the deterioration of his environment (through increased social functions in urban living, etc.), and to stresses imposed on his physical and mental state. The greatest danger arises from the fact that the physical and psychological limitations which become effective at the man-technology interface, are so poorly known. As René Dubos points out, the science of environmental biology for the human species has started only recently for military and space purposes. We do not know what degree of artificiality we can support in our environment, but some of the first results of the new discipline of neuropsychobiology seem to indicate that we are close to some absolute boundaries, or have already over-stepped them. The same may be suggested by mental sickness in urban environment, and by such phenomena as mass hysteria, increas-

ing violence, crime and other forms of alienation, and the increasingly neurotic behaviour found in the cities primarily among the young. Over the past millennia man has shown a remarkable capacity for adaptation to a more artificial world. But the question now is whether he can adapt further at the greatly increased rate of change and newly introduced aspects of an artificial world, and whether there are absolute limits to his adaptability.[2]

One may believe that useful functions for this subsystem may be defined, for example, in the following way: physical health, mental health, adaptation to physical environment, adaptation to psycho-social environment (which, of course, also belongs to the man-society subsystem), personal integrity (particularly privacy – in which area the invasion by technology may go much further than it has done so far, providing incentives for the development of such devices as personal information filters).

Failure to investigate and respect the boundaries set for the man-technology subsystem would lead in the most direct way to irrational response, and a dramatic aggravation of the crisis in the dynamically unstable overall system Nature–man–society–technology. It would also render totally ineffective the good, but isolated, planning in the society-technology subsystem which is coming into the spotlight today. The situation is particularly dangerous, since planning in the man-technology subsystem can hardly start in a serious way before much more basic research into the man-technology interface has enriched our information basis for this type of planning.

The vast spectrum of possibilities of engineering man through technology is only now becoming visible. With biology opening up the potential to change the biological foundation of man, we are given, in principle, a powerful means of changing the relationships within the overall system, for example by adapting man to specific natural or artificial environments, by eugenics, or by a considerable extension of his life span. On the other hand, we may also hope to find the means to counter the degeneration of the human species which, otherwise, would result from the elimination of natural selection processes for man. The basic potential and limitations of 'human engineering' in this direction also are only poorly known. Proceeding without sufficient information and without a good deal of wisdom may be disastrous, but so may be not proceeding at all. We have partly destroyed Nature and partly messed up society – we shall soon be able with equal ease to change or destroy the human species.

Apart from the three subsystems discussed above in which technology is involved directly as one of the two 'partners', there are three more sub-

[2] René Dubos, *So Human an Animal*, Scribner, New York 1968.

systems which do not include technology directly. They will be enumerated only briefly:

(4) *The Nature-man subsystem:* Much of the challenge to man to create an 'anthropomorphous' world reflects, of course, the potentials and limitations implicit in the Nature-technology subsystem. It should be noted, however, that the biological platform of man is just one aspect of his existence as a creature of Nature. Other aspects have to do with the overwhelmingly important psychological relationship of man with Nature, with the archetypal character of his thinking and his irrepressible sub-conscious image-forming (which suffers so badly under the imposition of images through technology), philosophy, religions, and the arts. Suffice it to say that 'human fulfilment', the value which is coming back to us today so powerfully, depends on the good 'management' of this subsystem more than on anything else, thereby also imposing severe limitations on the development of an artificial world through simple technology engineering.

(5) *The man-society subsystem:* The types of alienation which become visible in this subsystem frequently have their origins in technology introduced for the sake of the society-technology subsystem, especially in the context of urban living. The adaptation of and to the psycho-social environment of man – including all his psycho-social creations, such as political systems, nations, economy, etc. – here come into focus. Education is a particularly important function which also creates needs for new technological developments.

(6) *The Nature-society subsystem:* An important element here, which generates big demands on technology, is exploration, which could, ideally, provide the challenge of 'worthwhile adventure' which may become badly needed in a world going to higher degrees of automation and longer periods of leisure time. In this respect, the expensive space adventure has so far been improperly represented. Nigel Calder's[3] notion that the challenge of 'worthwhile adventure' may be found in the observation of unspoiled Nature, once technology (especially non-agricultural food production technology) generates the possibility of such land use, or better non-use, illustrates the wide margin of interaction with technology engineering. The 'biological landscape' function, discussed above under the Nature-technology subsystem, also impinges on the Nature-society subsystem, of course.

[3] Nigel Calder, *The Environment Game*, Secker and Warburg, London 1967.

THE VERTICAL AND HORIZONTAL INTEGRATION OF PLANNING –
TOWARDS INTEGRATIVE PLANNING FOR ECOLOGICAL
ENGINEERING

The formulation of planning objectives in terms of functions within
'bi-polar' subsystems gives us the possibility of controlling the variety of
feasible technologies by treating them as strategic alternatives to achieve
given functional objectives. This is not yet sufficient. It has already become
obvious that these functions interact with each other in a very complex
way, and that also the six 'bi-polar' subsystems interact with each other.
If we take, for example, the function of creating a 'biological landscape',
all other functions involving land use (food production, urbanisation,
transportation, etc.) are affected, and a good solution cannot be found by

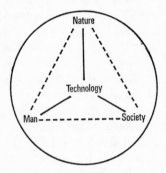

Fig. 3.3 Ecological engineering
establishes control over the total
integrated system; it deals with
the inter-relationships between
the functions defined within the
six subsystems.

satisfying only one isolated function. Also, functions may be 'mobile' over
time, especially in the society-technology subsystem.

Integrative planning, ideally, will attempt to go again to a higher level
of abstraction and to gain control over the interplay of the functions. The
first step leads to dealing with entire 'bi-polar' subsystems and an integrated approach to, say, the society-technology subsystem. Whereas this
is useful if we consider, for example, the planning of social change through
technology and of social institutions in an integrated way, we have still to
make a further step and integrate the functions that come under all subsystems. This is sketched in *Figure 3.3*, where a common planning and
control system deals with the six subsystems in an integral way. We could
call integrated action at this level 'planetary systems engineering', if this
would not suggest other meanings – but a proper name may also be found
in *'ecological engineering'* using the term ecology here for the totality of
the relationships in a viable system. The continuously changing images
guiding planning and control at this level are the discrete 'possible futures',
which Ozbekhan calls 'anticipations', Kahn 'alternative world futures' and
de Jouvenel 'futuribles'.

It must be said right here that we are far from being sufficiently advanced to have a planning methodology to deal with the total area embraced by 'ecological engineering'. Our capacity for systems analysis stops at single functions (such as food production on a world-wide scale), if it can handle them at all. However, the nearly unlimited capacity for systems simulation makes it possible to check the outcome of certain combinations of functions in the overall system, or of technologies within a function.

In other words, we are always to some extent capable of integrating our plans *vertically* – from technological options over functions to complete anticipations – and of checking the outcomes of certain combinations. But we are not capable of going beyond single functions in the systematic *horizontal* integration of our plans. This need not bother us if we could deal with the outcomes of specific technologies and the outcomes of specific strategies for the development of entire functions as if they were unambiguously defined building blocks. Unfortunately, this is not so. The outcomes of a selected technology, or a selected strategy for the development of a function, vary according to the other technologies, or strategies, introduced into the overall system, although in different degrees of sensitivity. There is no escape, at present, from using increasingly partial information as a base, the higher and the more integrative planning becomes. One may remember here, for example, the uncertainties introduced by the not very well known man-technology interface.

An added difficulty may arise when functions are not clearly definable over longer periods of time. Developments maturing in a farther time-distance might have to tie in with functions differing somewhat from those aimed at by developments which can be expected to impinge on these functions earlier. Especially in areas of very complex interaction between rapidly moving technologies, *'mobile' functions* may complicate forecasting and planning. For example, the future function of the 'home', and of urban development, in general, will depend to a large extent on the development of such functions as automation (people not going to work, but interacting from their homes by means of a computer console, etc.), communications (children and students learning at home via satellite communication, tele-shopping, etc.), and transportation.

Looking at the *vertical integration of technological forecasting and planning*, we have thus defined the following levels of objectives:

anticipations
('bi-polar' subsystems)
functions
technologies

Although objectives may be formulated for entire subsystems, they do not define a separate planning level in strict terms. Their introduction mainly serves a clearer formulation of functions. It would be artificial, for the interactions between functions at the next higher planning level, to exclude obvious inter-relationships between functions of different subsystems. With the present emphasis on the society-technology subsystem, we are used to dealing even with functions from the Nature-technology subsystem (e.g. environmental control) and the man-technology subsystem (e.g. noise control) as 'social' functions.

We have, in principle, to plan on the basis of feedback loops between a

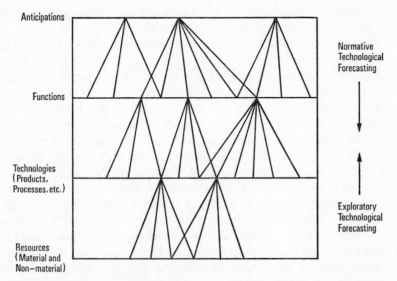

Fig 3.4 Targets for planning and forecasting. Each technology can, in general, be realised by different combinations of resources, and has a variety of outcomes; each functional objective has a variety of technological solutions. 'Commonalities' are indicated schematically. The three layers of the figure correspond to tactical, strategic, and policy planning. Technological forecasting implies simulation runs through every possible combination of unbroken vertical lines.

number of functions which may belong to different subsystems: some of these feedbacks between functions may be very strong and dominate the overall systems outcome of the introduction of a specific technology, and other feedback relations may hardly matter.

In short, it appears sufficient for most technological planning purposes to distinguish between three levels of objectives: technologies, functions, and anticipations. *Figure 3.4* gives a schematic picture of the scope of technological planning. A specific technological realisation (a product, process, or technological system) corresponds to a multitude of outcomes as well as, in general, to a multitude of resources to be employed for its realisation. Before deciding on the development of a specific technological realisation, one should be able to simulate any combination of links between the resources and the anticipation level – not just for one possible technological realisation, but also for all other technological realisations which are related to the same functions. Only in this way can the *technological decision-agenda*, the aim of technological forecasting at strategic planning level, become really informative as to the alternative decisions possible.

Clearly, such 'complete' simulation runs can be made only for technological developments which have reached an advanced stage. 'Incompletely filled' pages will therefore have to be added to the decision-agenda, which may nevertheless influence the decisions considerably.

Figure 3.5 attempts to depict schematically the vertical integration of technological planning in an industrial context. It shows the feedback cycles between the corporate environment – which will reflect mainly the society-technology subsystem – and the science and technology base at each corporate planning level, but also the triple feedback loop between technological forecasting and planning at the three corporate planning levels. Of particular importance, although less well developed as yet in terms of techniques and formal approaches, is the 'high coupling' between non-adjacent levels of objectives, e.g. between the functional decision-agenda and the basic scientific and technological resources. Few companies, however, already pay particular attention to such 'high coupling'.

A merely tactical (operational) approach to technological forecasting and planning would focus on technological objectives and correspond to simple (linear) product line development. The combination of the tactical with the strategic approach – bringing into play the full technological decision-agenda, i.e. the full spectrum of technological options – leads to a flexible technological approach in the framework of a specific isolated function which is assumed to be rigid. But only the addition of the policy level permits the rational setting of objectives which tie-in with the goals of 'ecological engineering' in a flexible framework of inter-related functions.

If we want to control the outcome of technology, we have to work back from an anticipated outcome. This is also, if we do an ideally good job, the only way to plan for consistent and reasonably stable futures. This is what we understand under *'actively shaping the future'*. Forecasting and planning provide the insight into the necessities and consequences of alternative decisions. They do not anticipate or favour a bias in the decisions.

The catalogue of available technological forecasting and planning methodology is rich in techniques pertaining to one planning level only (particularly the operational or tactical planning level, until recently the only generally recognised level), or to the interface of tactical and strategic planning. But it is relatively poor in techniques that attempt to span two or all three planning levels and which are of particular value for the vertical integration of technological planning.[4] The most elaborate planning technique that has been applied to the vertical integration of planning emphasises the nature of integrative planning as planning in a time continuum, not in a rigid time-frame: The Planning-Programming-Budgeting

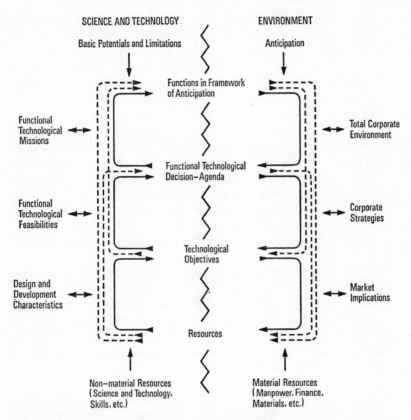

Fig 3.5 Integrative technological planning in a corporate framework. The three main *feedback* loops correspond to policy, strategic, and operational planning; the dotted lines indicate the possibility and desirability of 'high coupling'. Planning at every level has a system-wide scope: anticipations, functions, and technological systems, respectively. The characteristic inputs of technological forecasting are indicated for every level; on the environment side, they are enriched by inputs from non-technological forecasting.

System (PPBS) is a scheme which makes it possible to follow dynamic programme development aimed at functional objectives.

The effective vertical integration of technological planning in an institutional, national, or international framework will depend chiefly on developments in information technology, permitting the organisation of

[4] See Chapters 5 and 6 of this volume.

vast input information in the form of environmental information systems,[5] the design and storage of models of the present and the future as well as cause-effect relationships, and the simulation of large numbers of alternative courses of action. There can be little doubt that the development of information technology, in full swing now, will make this feasible.

Looking at the *horizontal integration of technological planning*, aiming at system-wide planning at each of the planning levels discussed above, we encounter difficulties in two directions: (a) the vast number of inter-relationships to be taken into account in large systems imposes a limit on our capacity to plan for such systems integrally; and (b) we quickly run into difficulties concerning the measurability, i.e. the quantitative comparability, of the outcomes of technology.

Let us deal with the second problem first. It is the one which ought to be given top priority in the development of planning methodology. A useful approach has so far been developed in only a narrow sector of the society-technology subsystem, namely for defence technology, which has least to do with the 'quality of life' we are attempting to plan for. This is the cost/effectiveness approach, which aims at the quantification of the effectiveness which a specific technological realisation has (or is predicted to have) in a large system, usually defined by one of the functional 'program categories' of the PPBS approach.

Numerous approaches exist to simple cost/benefit analysis; they aim at quantifying the direct and predominant benefit to be derived from a planned technological realisation. These benefits may be expressed in terms of technological capabilities such as speed, strength, temperature resistance, and the like, or – as is widely done in an industrial and business context – as economic benefits, usually expressed in terms of return on investment, or – with discounted cash flow calculations taking into account the 'time is money' factor – as present net value.

What is badly missing, however, is a *generalised cost/effectiveness approach in two stages:* the first stage would permit us to compare the outcomes of specific technologies within a functional system (such as 'urban transporation', where a particularly great variety of non-linear changes in the form of new technologies ought to be introduced into the horizontally integrated planning of this function). The second stage would then allow us to compare the effectiveness of plans for entire functions in relation to plans for other functions – in short, aid the effective integration of functions in the framework of anticipations. This second stage may still be relatively far off. But we can hope to develop the first stage within the next few years, if we recognise its importance.

A few modest starts have been made in this direction, for example by introducing weight factors pertaining to such elements as comfort and

[5] Theodore J. Rubin, 'Toward Information Systems for Environmental Forecasting', in E. Jantsch (ed.), *Perspectives of Planning*, OECD, Paris 1969.

status in comparative studies of private and mass transport, etc. But these are relatively poor beginnings. We ought to know how much it is worth to have the air pollution level of a specific city reduced by ten, twenty, or fifty per cent, of having noise level reduced by a specific fraction, and personal mobility increased at the cost of independence. We ought even to be able to quantify such secondary phenomena as the absence of birds from polluted cities, and the variation of human life expectance with the stress imposed by urban life. There will be a wide variety of opinion on these measures of the quality of life. But we cannot hope to plan for the quality of life in other than intuitive terms if we do not attempt to define measures (which, of course, may be derived through public consensus, and refined weighting of deviations from it).

The purely economic thinking and crude cost/benefit approach is characteristic of the autonomous development of technology engineering and a belief in the sequentiality of events such as the introduction of specific technologies. It does not even look at the economic consequences of the endless technology/counter-technology sequence that follows to counteract the harmful impact of technology engineering on Nature, human, and social engineering. It obstructs our view of the quality of life.

In dealing with technology in the 'bi-polar' subsystems, economic analysis becomes a part of cost/effectiveness. Desalinated water costs ought not to be compared with the costs of present natural water resources, if it is a question of making a desert habitable, or of preventing the development of arid zones where the ground water level has been lowered, or the desertion of industrialised areas which lack water for further industrial development. The unit and operating costs of non-polluting cars ought not to be compared with those of internal combustion engine cars, or the price of lead-free gasoline with that of lead-containing gasoline (a technological problem already solved). Single Cell Protein produced on substrates such as petroleum and organic wastes ought not to be compared with the costs of skim milk powder, which cannot solve the world food problem to which Single Cell Protein could make a decisive contribution. It cannot be seen, how, for example, the function-oriented Planning-Programming-Budgeting System can ever work in civilian government areas if no social cost/effectiveness technique is soon developed.

There is also a special role for governments here: adapting cost/effective solutions to an economically oriented environment, for example, by taxing the difference away so that the economic characteristics are at least equal, if not advantageous, for the socially better solution.

The other difficulty encountered in the horizontal integration of technological forecasting and planning, namely the limitations on *system-wide planning* due to the number of inter-relationships to be taken into account, becomes visible in three ways:

(a) The number of interacting elements, and their individual variation, may
    be large. For example, the number of potential basic technologies and
    their variations may render the problem of evaluating every possible
    combination sizeable, or even excessive.
(b) The part of the system which is of concern for a specific planning enter-
    prise may vary according to purpose. The 'food production' function
    may be relatively simple for the individual farmer who has to make the
    choice between different plant and animal species and their combina-
    tions, but it may be much more complicated for food industry with a
    developed market system and it may become of hardly manageable
    size if seen in the light of the world food problem. Many of the most
    important functions within the society-technology and the Nature-
    technology subsystems assume world-wide dimensions, and require a
    concerted approach of the whole world, or at least the advanced
    countries.[6]
(c) The possibilities of non-linear change are restricted not only by the
    systems boundaries, but also by the inertia inherent in the various
    dynamic elements of the system, especially in complex social systems.
    We cannot change instantly either the direction or the momentum of a
    dynamic system – only the curvature of its movement. The ultimate
    solution of the world food problem (if population growth cannot be
    drastically influenced in the immediate future) will require the develop-
    ment of non-agricultural food production technology. This develop-
    ment has to be achieved, and introduced, between now and the year
    2000. At the same time, however, agricultural productivity has to be
    increased by every possible means, even if it should become clear that
    part of agriculture will subsequently be abandoned. This type of
    simultaneous management, possibly assuming the character of crisis
    management in many functions of the society-technology and Nature-
    technology subsystems, will dominate in the decades to come.

As was pointed out at the beginning, the horizontal integration of
technological forecasting and planning implies integrated planning of the
vertical and the horizontal technology transfer, of technology and its
outcome. In the future, it will mean to an ever increasing extent the planning
of systemic change (in particular, social change). This implies not only
planning for new applications and new services, but also planning of entire
systems involving technology, of new forms of organisation and new
institutions. Computer technology provides a good example: having
started with hardware development and software development (already of
equal market importance) the planning tasks in this area now extend to
the integral planning of organisations along with hardware and software
(e.g. in management information systems), and will soon incorporate the

[6] See Chapter 14 of this volume.

planning of new types of institutions and inter-institutional structures. In this context, it becomes of particular importance to plan for the *rate of change* and for the introduction of technology to fit a rhythm which is in step with the systems dynamics. The methodology for achieving this task is still in a very primitive stage.

The present state of methodological development leaves us still pretty far from the target of planning for 'ecological engineering', or for entire anticipations. However, quite a number of inroads will permit at least partial studies of dominant relationships between functions, and therefore of aspects of 'possible futures'. We should still need the general cost/effectiveness approach permitting us to quantify the merits of strategies for entire functions. But, most of all, we should need much more knowledge about the relationships at work in the man-technology subsystem, from where the most dangerous threat to our naive 'ecological engineering' attempts will emanate.

The task of integrative technological planning has recently been evoked and summarised by the Vice-President in charge of research of the Bell Telephone Laboratories:[7] 'The greatest technical innovation of the future will involve whole systems of transportation, construction, public health, agriculture, defence, communications, and so forth. The basic scientific factors that underlie the elements of the system will become ever more important. . . . Basic science will lead us to new realms of innovations in which the life sciences, the physical sciences and the behavioural sciences will be combined in new ways of learning and in new forms of society. The time has come, with the aid of simulation and the statistical capabilities of the digital computer, to seek many more common languages and common concepts for the joint systems of man and Nature.'

[7] William O. Baker, 'Broad Base of Science', in series Innovation: The Force Behind Man's March into the Future, *New York Times*, 8 January 1968.

4. Technological Forecasting in Corporate Planning

NON-DETERMINISTIC FORECASTING

THE widespread introduction of industrial technological forecasting, starting in the first half of the 1960s in the United States, followed the breakthrough of corporate long-range planning with a phaseshift of approximately one decade. This relatively close succession of significant innovations in corporate management did not come about accidentally. Corporate long-range planning is essentially a *strategic concept*, dealing with alternative options, and so is technological forecasting. At the same time, corporate long-range planning adds a *qualitative* dimension to the predominantly quantitative framework of extrapolative planning. Technological forecasting, in attempting to bring potential future technologies into focus, aims primarily at targets which differ qualitatively from the present state. Thus, the development of technological forecasting as an integral element of corporate long-range planning was enforced by the evolution of corporate planning in areas marked by rapid technological change.

It is obvious that the extension of corporate planning beyond its original framework of extrapolative planning marks the *transition from a deterministic to a non-deterministic planning model*. This is the decisive aspect to be watched for the proper introduction of technological forecasting into a modern corporate planning concept.

Technological forecasting, like corporate planning, originally developed along the lines of a deterministic approach to planning. It adopted the characteristic principles of linearity and sequentiality, to which the rather indiscriminate use of time-series relating to single technical parameters still bears testimony.

In this primitive form, technological forecasting provided certain inputs to extrapolative planning, but did not interact with other elements of the planning process. It was, in its extreme interpretation, regarded as an objective source of truth about the future, external to planning and human action.

These nowadays obsolete notions also appear to underlie the still wide-spread misunderstanding which believes in separating technological forecasting from planning–forecasting as the supposedly objective, planning as the subjective element. In reality, technological forecasting forms *an integral part of the full corporate planning process.*

For modern corporate planning, and in particular for technological forward planning, the purely extrapolative forecasting and planning approach yields nothing but feasibility indications for targets as well as for the pace at which they may be attained. The possibility of extrapolating technological capabilities, e.g. by means of envelope curve extrapolation[1] permits the use of extrapolative tools for strategic long-range planning beyond existing and emerging technologies. It provides a means to 'look further than one can see', to apply Churchill's paradox. Used wisely, this may become of particular value.

In general, however, the evolution of corporate planning, from predominantly extrapolative to strategic and then to policy planning, implies an ever decreasing applicability of linear and sequential modes of forecasting. Not only will qualitative planning and the growing importance attached to value dynamics override the inertia inherent in linear developments. Simultaneous planning in the framework of large functional systems, and eventually of entire anticipations ('possible futures') will also render the extrapolation of progress in single technological parameters or capabilities increasingly meaningless. All will depend on the interaction between a large number of technological and non-technological parameters, and on the qualitative goals sought.

Technologies are developed to match needs. The deterministic approach to planning inherently assumed that specific technologies corresponded to specific needs in an unambiguous way. All that needed to be done for technological forecasting in this framework of sequential problem-solving was to identify needs and technologies and match them. The additional assumption of linearity in extrapolative forecasting tacitly assumed that linear trends of technological parameters or capabilities automatically benefited goals of the future in the same way as they had benefited goals of the past. This imposition of pseudo-goals arising from the inertia of technological development, and not selected for their merits in a future context, has been rightly condemned by Galbraith.[2]

In contrast to this, advanced corporate planning in technological areas is now based on the recognition of the following facts:

—A technological development, in general, has a variety of possible outcomes;
—A problem or a set of problems, in general, have a variety of technological solutions, or inputs;

[1] See Chapter 6 of this volume.
[2] John K. Galbraith, *The New Industrial State*, Houghton-Mifflin, Boston 1967.

—Technological developments form part of a system and have to satisfy simultaneous strategies with a system-wide scope.

Exploratory forecasting, which starts from today's level of knowledge and explores future feasibilities and probabilities, has to match *normative forecasting,* which implies the delineation of goals of the future and their translation into missions and tasks for scientific and technological development. Only the combination of both in a 'bi-polar' view leads to technological forecasting fitting into a non-deterministic framework; exploratory forecasting alone is a deterministic concept.

The inherent characteristics of technological development enhance the role of normative thinking in a significant way. Almost all technological innovations can be traced back to a clear need formulation.[3] The enumeration of opportunities alone normally does not trigger action. Modern corporate management takes this into account by devoting increasing attention to the problem of translating corporate objectives all the way down the hierarchic ladder to stimulate the creativity of people at all levels. More recently, emphasis is being placed on linking corporate objectives to social goals, thus starting the normative chain at the highest level. 'The corporate leader who does not try to conduct his company so as to instill pride in his people is doomed these days' (Joseph Wilson).

A complete technological forecasting exercise, on this basis, always constitutes an *iterative process between exploratory and normative forecasting.* This process, it should be noted, does not match rigid sets of technological opportunities with equally rigid sets of objectives, ordering them in such a way that 'the right key fits the right hole'. It constitutes a feedback cycle in which both opportunities and objectives are treated as adaptive inputs. The search for new and more suitable opportunities goes hand in hand with the adaptation of goals to technological feasibilities and higher probabilities. Technological forecasting in the framework of integrative planning includes feedback processes of even higher complexity.

THE FORECASTING TIME-FRAME

In the framework of linear planning, technological forecasting focussed primarily on the recognition of future technological realisations in concrete, descriptive terms. The point might be made that the scientific principles

---

[3] This is borne out, in particular, by the following evaluations of R&D events contributing to the realisation of technological systems: Chalmers Sherwin and Raymond S. Isenson, *Project Hindsight,* Interim Report, Department of Defense, Washington, D.C., October 1966; Arthur D. Little, Inc., *Patterns and Problems of Technical Innovation in Industry,* Cambridge, Massachusetts 1963; and an unpublished study by the TEMPO Center for Advanced Studies, Santa Barbara, California.

known today, with relatively few exceptions, constituted the seeds for technological innovations over a period of not more than *some 15 years*. Moreover, it was considered superfluous to go beyond this time-distance, since good forecasting would guarantee a 'surprise-free' period sufficiently long not to worry about the opportunities presented by new scientific discoveries.

This limitation in outlook still applies to technological forecasting for the purposes of *operational planning*. Only at this lowest of the three planning levels is it the task of technological forecasting to envisage concrete technologies and technological systems. Normally, one will keep the options open until the decision about a certain line of technological development has to be taken. The time-span of 5 to 10 years' development generally encountered in complex development is frequently also adopted as the time-frame for technological forecasting in an operational planning context.

It is evident and essential to note that no implicit time restriction applies to the forecasting of technological capabilities or basic potentials and limitations, as required as an input to strategic planning. It is easy, for example, to forecast the technological capabilities which can be generated by nuclear rocket propulsion. Because the lightest possible propellant gas, hydrogen, can be used, the specific impulse (or, in other words, the nozzle exit velocity) can be tripled as compared to the lightest combustion exhaust gases achieved in chemical propulsion. This advantage can be readily translated into capabilities in terms of greatly extended mission ranges. In the same way, the capabilities created by electric rocket propulsion (and the restrictions, such as those set by the Sun's and the Earth's gravity) can be readily forecast. We can even forecast, in terms of technological and mission capabilities, space systems making use of the ultimate in specific impulse, namely photon propulsion, although this potential may really be very far-off.

The structuring of thinking in accordance with long-range and ultimate goals, which constitutes the essence of *normative planning*, also widens the time horizon of technological forecasting. The goals for the future, which are usually very different from the goals applying at present, provide an outlook that may require a technological decision-agenda quite distinct from one appearing satisfactory for a shorter time-frame.

It is not surprising to find that those branches of industry which, so far, have pioneered the application of strategic and normative planning in their corporate thinking, practise technological forecasting in a time-frame which extends up to 50 years, and sometimes beyond (*see Table 4.1*).

These are branches which are conscious of the social implications of the technologies they develop, and therefore try to anticipate social goals of the future. Technical companies of an innovating type, such as the aerospace, electronics, and chemical industries, usually apply technological

	Formal	Informal
'Look-out' institutions (evaluating overall future systems), 'social technology' (technology with considerable impact on society, such as automation), natural resources		up to 50 (*sometimes even beyond*)
Industry in 'social technology' areas (big oil companies, communications sector)	5–10	30–50
Space programme (NASA)	10–20	20–30 (*sometimes beyond*)
Defence	7–10	20–25 (*sometimes up to 40*)
National economy (French Plan)	5	20–25
National goals (American civilian government)	5–6	10–30
Industry of an innovating type (electronics, aerospace, chemistry), with a corporate long-range planning function	5–10	10–20
Consumer-type industry	3–5	5–10

Table 4.1. Typical Time Scales of Forecasting (in years).

forecasting to a time-distance of 10 to 20 years, whereas consumer-type industry is satisfied with 5 to 10 years. These indications refer much more to the emphasis placed on strategic and normative planning than on a difference in 'visibility' of the technological developments and their outcomes. In general, operational technological plans, integrated in the formal corporate long-range plan, hardly ever go beyond a period of 10 years.

When then is the limit for normative planning? In many contexts, it appears to be set by the dynamics of the global system of human society. The disequilibria and dynamic instabilities developing in this system at present, namely the 'gaps' between the developed and the less developed countries, the unfinished trend towards world unification, and the inertia of the population explosion, can possibly be brought under control not before 100 or 150 years have elapsed. If, in normative planning, a new situation of global stability is envisaged it is a time-frame of approximately this length which is implicitly applied. On the other hand, if the aim is just vaguely formulated in terms of an improvement over the present situation, the goal of this improvement is not clear.

It does not appear, therefore, as utopian if a complete corporate planning framework includes technological forecasting which aims at goals 100 or 150 years in the future – although it will certainly not be able to specify by what means these goals will be attained. But it will attempt to verify whether the technological decision-agenda resulting from the application of existing or imminent technology is sufficient to reach those distant goals. If this is not so, technological forecasting will stimulate the search for new technologies and for new types of fundamental research.

THE UNFOLDING SCOPE OF TECHNOLOGICAL FORECASTING IN CORPORATE PLANNING

As stated above, the task of technological forecasting is not to predict technological realisations of the future. The inputs to strategic and normative planning do not generally include forecasts in terms of concrete technological systems and their design elements. But it must also be noted that the scope of technological forecasting in the various stages of corporate planning differs widely. This gives rise to quite different approaches, depending on the planning stage, which are taken simultaneously in a complete framework of corporate planning.

Figure 4.1 attempts to follow a technological innovation through its various stages identified by the RDT&E phases:[4] pre-discovery, discovery, creation, substantiation, development, advanced engineering, application and service engineering.

This succession of RDT&E phases provides a much clearer picture of technological development than the widespread use of the ill-defined notions of fundamental research, applied research, and development – which, in addition, do not form an unambiguous succession of the actual work phases, but overlap and supplement each other particularly in the decisive stages of creation and substantiation.

It can be readily seen in *Figure 4.1* that exploratory forecasting, starting from the known science and technology base, and normative forecasting, starting from the environment, cannot be matched directly and in explicit terms until relatively late in succession of RDT&E phases. Although some tentative forecasts, made between the creation and the substantiation phases, may attempt to anticipate possible systems designs in descriptive terms and possible market impacts, these forecasts may still undergo considerable modification on the basis of the results obtained in the substantiation phase. It is only before the beginning of full-scale

[4] RDT&E is the commonly used abbreviation for 'research, development, testing, and engineering' comprising all scientific-technical aspects of the technological innovation process (whereas the traditional notion of R&D, i.e. research and development only, is more restricted). These stages were adapted from a scheme proposed by the Stanford Research Institute.

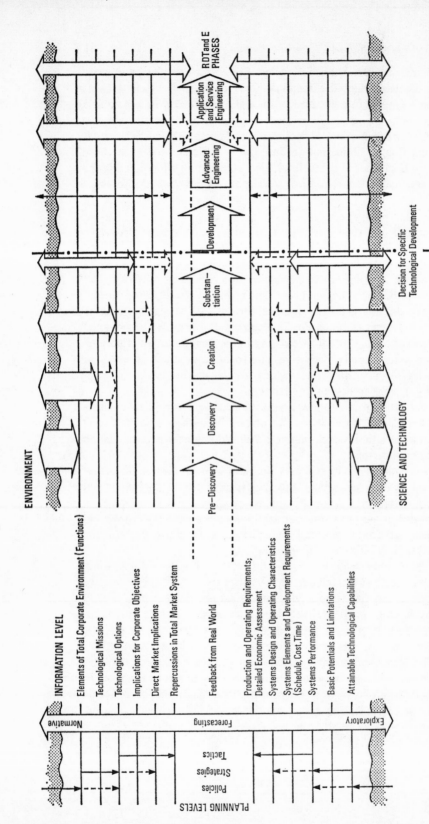

Fig 4.1 *Technological forecasting stages in relation to the RDT&E phases of a typical technological innovation.* The formal scope of each technological forecasting stage is the forecasting of the outcome of the next RDT&E phase, the informal scope (dotted arrows) extends further. Six forecasting stages and one evaluation stage may be distinguished. The thickness of the arrows indicates schematically the width of forecasting, narrowing down in the course of vertical technology transfer from the simultaneous consideration of a large number of alternatives to the decision for one specific development line, and subsequently widening again in considering alternatives of horizontal technology transfer (applications and services).

development work that such a forecast enters formal planning and leads to a decision for a particular development line. Only then can needs and opportunities be matched in a complete forecasting cycle. *Figure 4.1* tries to express this by using forecasting arrows of different thickness, indicating that, as long as exploratory and normative forecasting do not clearly 'see' each other, a number of alternative possibilities has to be taken into account simultaneously. After 'vertical' development for scientific principles up to technical realisations turns to 'horizontal' development adapting the new product or system to various applications and services, forecasting again includes various alternatives simultaneously.

The forecasting arrows are depicted as pointing in both directions, thereby indicating that all forecasting is a continuous feedback process. Norms derived from the environment, and technological feasibilities derived from the known science and technology base, are continuously modified and updated against these ever-changing backgrounds.

Figure 4.1 also indicates the structure of one of the most important tasks of technological forecasting, namely *to improve the coupling between successive RDT&E phases*. By anticipating the possible results of each work phase, technological forecasting stimulates and guides work to be undertaken in the successive phase. No fewer than six different coupling functions can be read from *Figure 4.1*. Of particular importance is the coupling between the discovery, creation, substantiation, and development phases – in vaguer terms the famous coupling between science and technology, which is blamed today for Europe's lagging behind in some technological development areas in spite of the sound scientific basis available.[5]

One would think that, ideally, *there ought to be a technological forecasting exercise at each of the six 'coupling joints'*. This is indeed the goal towards which the detailed structure of technological forecasting is developing today in advanced thinking corporations.[6] The first two forecasting stages may be handled in an informal way where no formal policy planning has yet been established. Of decisive importance for successful product development are the forecasting stages between creation and substantiation, and between substantiation and development – in other words during the transition period from strategic to operational planning.

A case study of the development of the TH System,[7] a long-distance microwave relay system, at Bell Telephone Laboratories, from the start in 1952 to manufacturing in 1958 mentions two formal forecasting exercises, carried out by the Systems Engineering people. In 1952, a clear recognition of the technological mission and alternative development lines, and of feasible and desirable systems performance, as well as a first estimate of

[5] See also Chapter 7 of this volume.
[6] See Chapter 10 of this volume.
[7] Thomas A. Marschak, 'Strategy and Organisation in a System Development Project', in *The Rate and Direction of Inventive Activity*, Princeton Univ. Press, Princeton 1962.

economic demand; in 1954, a second forecast including development time and costs, development schedule, and detailed systems objectives. The choice between alternative major systems characteristics was made on the basis of this second forecast; the decision was for the TH System and detailed operational systems planning started.

It is important to note that, as exploratory and normative forecasting approach each other, the *forecasting scope* does not shift from level to level, but expands to include more and more information levels. If the first forecasting stages may be compared to visualising vague mirror images of only the most basic formulations of needs and potentialities, the succeeding stages build more and more firmly on this basis.

Technological forecasting, if practised correctly in the framework of advanced corporate planning, is emphasising *'high coupling' within the iterative forecasting cycle itself*, meaning that gradually each information level is matched with all other levels. For example, forecast systems design elements are matched not only against direct market implications (economic demand, price, etc.), which would constitute 'low coupling' but also against the norms derived on the level of technological missions, at the level of the elements of total corporate environment, as well as against attainable technological capabilities and basic potentials and limitations, and so forth.

TECHNOLOGICAL FORECASTING AT POLICY-PLANNING LEVEL

The scope of technological forecasting is different for each of the three planning levels corresponding to the advanced corporate planning framework. Instead of attempting to formulate a generally applicable and necessarily vague definition, it is more useful to distinguish between technological forecasting as applied to planning at these three levels.

The scope of technological forecasting at policy-planning level is the clarification of scientific-technological elements determining the future boundary conditions for corporate development. In more concrete terms, technological forecasting at this level focusses on basic scientific-technological potentialities and limitations as well as on their conceivable ultimate outcomes in a large systems context.

Non-linear planning at policy level deals, at one and the same time, with problems for which the inherent freedom of choice is very narrow and others for which it is very wide. The tools which serve the purposes of technological forecasting at this level range from quasi-linear extrapolation to the almost uninhibited use of intuitive imagination.

A small degree of freedom is encountered when dealing with large systems of the corporate environment, which may range from market systems and industrial sectors to political and societal systems of national and even global dimensions. Large dynamic systems, especially if the aspects of social dynamics dominate, imply a relatively high inertia of macroprocesses. This becomes visible, for example, in demographic, political, and macroeconomic trends. Of particular significance is the inertia inherent in institutions of all kinds.

Industry is itself such an institution and corporate development is restricted by the feasible degree of internal and external flexibility, including vested interests, qualitative and quantitative aspects of manpower, production capacity, marketing, and so forth.

A large degree of freedom is encountered in the adaptation of corporate policies – including their scientific-technological elements – to the environment. The multitude of technological solutions contributing to a specific goal, and the multitude of possible outcomes of each technological development in a large systems context, guarantee the possibility of flexible choice, including, in particular, non-linear changes in technological development.

Industry may be viewed as an emerging 'planner for society' and is thus assuming an active and decisive role in *'shaping the future'*.[8] In fact, it is the adoption of this basic attitude which, more than anything else, stimulates the introduction and application of the advanced corporate planning framework, and of elaborate technological forecasting, in industry today.

The idea of studying, in the framework of independent national and international institutions, functional systems as parts of complete *anticipations* (intellectually constructed models of alternative possible futures) is therefore of the greatest interest for industry. In the United States, the National Industrial Conference Board has recognised this and helped to create the 'Institute of the Future'.

Forecasting and planning at policy level face particularly difficult problems since, to a considerable extent, they deal with *value dynamics*. There are no infallible ways to judge the future merits of certain goals, or of outcomes of technological development. Nevertheless improved knowledge of the mutual consequence of the adoption of specific goals and of decisions for action in the present can be achieved through the simulation of planning alternatives in a feedback cycle. To a certain extent, this will already make it possible to distinguish between 'good' and 'bad' alternatives.

A special problem for technological forecasting for the purposes of industrial policy-planning concerns *future changes of inter-industry structures*. Apart from the acquisition policies of financial holding companies and diversification on the basis of special skills, there are at least three effects which are directly related to technological development:[9]

---

[8] See Chapters 12 and 13 of this volume.
[9] See also Chapter 8 of this volume.

—Diversification through 'organic' forward and backward integration to form integral functional chains (exemplified by the Esso/Nestlé enterprise for the development of Single Cell Protein on petroleum basis – the functional chain here covers the whole range from petroleum prospection to finished food stuffs).

—The 'invasion' of one industrial sector by another on the basis of technological innovation (exemplified by the invasion of the textile by the chemical sector, the chemical and pharmaceutical sectors by the petroleum companies, etc.).

—The 'blurring' of sectoral boundaries due to technological developments pushed to their extremes (exemplified by the vanishing distinction between electronic components and systems in the era of integrated circuits, and between airframe and propulsion in the era of supersonic and future hypersonic aircraft).

Figure 4.1 graphically shows the central problem of technological forecasting in the early RDT&E phases, when policy-planning is still isolated, without the benefits of elements resulting from strategic and tactical planning. It establishes, over a wide gap, *a dialogue between functional targets* on the environment side, *and basic potentialities and limitations* on the science and technology side. Thereby, technological forecasting stimulates in a significant way *fundamental research*, at the same time providing broad guidelines for it.[10]

For normative forecasting at policy-planning level a special formal approach is coming into use – the *Environmental Information System* (EIS) comprising data input and storage techniques as well as the processing of data to link them to corporate objectives.[11]

TECHNOLOGICAL FORECASTING AT STRATEGIC PLANNING LEVEL

The core of technological forecasting is situated at this level. It is here that the nature and impact of future technological options become discernible in their profiles, that the decisions leading to the adoption of specific technological development lines are prepared, and that imaginative forecasting is able to influence the course of social and economic macroprocesses most effectively.

The scope of technological forecasting at strategic planning level is the recognition and comparative evaluation of alternative technological options, or in other words, the preparation of the technological decision-agenda. The focus is not on descriptive forecasts in terms of systems design, e.g. how a new type of machine may look or work, but on assessing feasible systems performance in the light of attainable technological capabilities, and on relating technological options to functional missions. Nevertheless, as

[10] See Chapter 7 of this volume.
[11] Theodore J. Rubin, 'Toward Information Systems for Environmental Forecasting', in E. Jantsch (ed.), *Perspectives of Planning*, OECD, Paris 1969.

indicated in *Figure 4.1*, forecasting attempts to penetrate, to some extent, into areas which are to be clarified by operational planning.

The most important task for strategic planning is the elucidation of *cause/effect relationships over long periods of time*. The human mind, intuitively, relates causes and effects which succeed each other within relatively short time-spans. This implies the danger that incorrect or artificial relationships are established which tend to blur the issues of corporate long-range planning and replace them by a mosaic of short-range relationships.

The *large systems aspect* and the search for *simultaneous* strategies satisfying the boundary conditions and a multitude of systems requirements, constitute another basic task for strategic planning and forecasting. Strategies can, of course, aim at different systems, such as those formed by markets, resources, special skills, financial compounds, and functional areas.

It is most important to note that planning with a qualitative scope in a societal framework – the scope emerging now for advanced corporate planning in industry – has to be based on *functions*. Forecasting at strategic level evaluates technological missions and alternative options in the framework of broad functions, i.e. societal need categories.[12]

A third important aspect of forecasting at strategic level is the deliberate and unbiased introduction of *non-linear planning* alternatives. All forecasting techniques applicable at strategic level incorporate this element of non-linearity, and some of them are designed with this particular aim in mind, for example, envelope curves, morphological research, and contextual mapping. All of them, however, have some inherent limitations which give rise to 'high-order' linearities, which also have to be recognised and carefully avoided in sophisticated strategic planning.

It is significant that the 'low coupling' approaches in the forecasting cycle, e.g. techniques relating development requirements to market impact, are increasingly making use of strategic aspects. If resource allocation and priority ranking procedures on the basis of simple operations research formulae aim at maximising returns on investment (or 'present net values' in a discounted cash flow framework), their results are over-ruled by priorities derived from simple decision theory formulae which place emphasis on the strategic impact of certain development. This means that strategic thinking increasingly dominates also those RDT&E phases at which operational planning is starting.

TECHNOLOGICAL FORECASTING AT OPERATIONAL PLANNING LEVEL

If normative and strategic planning are still somewhat aloof and keep sufficient distance from reality to maintain a capability of flexibly

[12] See Chapter 8 of this volume.

choosing between alternative options – attempting to 'ride' the future reality, not be 'ridden' by it – tactical or operational planning is the level at which these optional concepts have to harden into reality in a process that involves the adaptation of both the plan and the factors of reality to each other. In technological forward planning, it is the results obtained at this level which enable the planner to make the decision for specific technological objectives and developments.

The scope of technological forecasting at tactical planning level is the probabilistic assessment of future technology transfer. Both vertical and horizontal technology transfer are considered here, implying that the forecast penetrates beyond the design and performance characteristics of a specific technological system to its utilisation in the context of different applications and services, repercussions in the market, system and implications for developments of social, technological or other nature.

Operational planning may also include a contingency approach, e.g. the various *network planning techniques*, which may be considerably enriched by technological forecasts related to critical 'obstacles' and decision points.

A particular aspect of technological forecasting at this level concerns its contributions of extrapolative planning, or linear planning in general. Only in this narrow framework can deterministic planning be put to use. Nevertheless, the skilful application of strong normative forecasting, in combinations with the purposeful assessment of absolute or practical limits, already implies the possibility of gaining decisive advantages over the competitors by developing an attitude of *'leading the main-stream'* by aiming at a steeper gradient of performance over time. This is the attitude of aggressive and alert technical corporations, aiming at high-risk profit as they dominate the market today in the United States.[13]

A number of techniques for *economic analysis*, the more sophisticated ones applying discounted cash flow calculations and optimisation by simple operations research, provides the 'bonding' between technological and market forecasting.

INTEGRATIVE CORPORATE PLANNING AND TECHNOLOGICAL FORECASTING

Integrative planning is planning cutting through two or more of the following dimensions: social goals and values, institutions and social structures, technology and society, population and material resources, physical environment.[14] Corporate planning within the framework out-

[13] See Chapter 8 of this volume for the three basic dynamic attitudes of industry.
[14] See Chapter 3 of this volume.

lined above is inherently integrative planning and technological forecasting contributing to it has to deal explicitly with all the above dimensions.

In Western civilisation, technology has become the most important factor in social change and substitution. The uncontrolled development of technology leads to metastable solutions and disequilibria, instabilities in the dynamic system of society, waste, and uncontrolled or cancerous change.

Within the corporate framework, the uncontrolled development of technology may mean, in particular:

—Technological feasibilities, when recognised, are turned into reality; 'can' is understood as 'ought';
—Market demand is followed 'blindly';
—Product lines are pushed to the extreme;
—Demand is created artificially for the sake of specific scientific and technological developments, or to maintain certain skills and capabilities.

In general, uncontrolled development occurs where a deterministic attitude is taken, in particular the attitude of linear and sequential planning.

Forrester[15] has shown that the simplest systems of real interest to the manager begin at fifth order and that most of the actual systems encountered in corporate management may be situated between tenth and hundredth order. At the same time, corporate systems have to admit positive feedback, a high degree of non-linearity, and multiplicity of loops. The same development may thus be expected to happen as in engineering when it reached a degree of complexity that outgrew conventional mathematics, namely the application of analogue and digital computers to replace analytical solutions of complex problems by simulation. 'Rather than naively oversimplifying the real world to fit within the limited power of mathematics, the engineer turned to the simulation of large numbers of special cases. From these special cases he can generalise, within limited regions, on the character of the system under study.'

The aim of corporate planning is therefore to determine 'good' strategies by selecting them from as broad a basis of alternatives as possible. Technological forecasting, especially at the strategic planning level, is called upon to enrich this basis for selection. At the same time, the systematic evaluation of a lage number of alternatives – implying the iteration between environment and the science and technology base by means of multiple simulation of the consequences inherent in the selection of any one technological option – underlines the importance of modern

[15] Jay W. Forrester, 'Common Foundations Underlying Engineering and Management', *IEEE Spectrum*, September 1964; and *The Structure Underlying Management Processes*, The Academy of Management Meeting, 30 December 1964, Chicago, Illinois.

information technology for the tasks of technological forecasting. One may expect that technological forecasting will be incorporated into future complex *management information systems*.

It should also be noted that, in a non-deterministic framework, technological forecasting forms an *integral element of planning*, not an 'external' contribution to it. The feedback cycle between the exploratory and the normative approach in non-deterministic forecasting actually implies a feedback relationship between forecasting and the design of concrete plans. To the extent that a technological forecast influences the design of the plan, this plan also influences technological forecasts. This, of course, is irreconcilable with the older notion of technological forecasting as some source of absolute truth.

It is of the utmost importance to understand that both *technological forecasts and plans are adaptive to each other*, and to make use of this flexibility in corporate planning. Running through the simulation of a number of preconceived alternatives will normally lead to second and further simulation runs 'testing' modified and adapted alternatives in technological forecasting and plan design.

Forrester[16] considers the future form of management as 'enterprise engineering' and points out that of the possible contributions which technical engineering could make to it, the following stand out:

(1) The concept of designing a system.
(2) The principles of feedback control.
(3) The clear distinction between policy making and decision making.
(4) The low cost of electronic communication and logic.
(5) The substitution of simulation for analytical solutions.

Technological forecasting may be considered as a tool of enterprise engineering involving, ideally, all these principles. The feedback character of technological forecasting, the need to employ simulation, and the implications of computer use, have all been brought out clearly above.

In analogy to technical engineering, Forrester envisages two kinds of engineers also for 'enterprise engineering': the *stationary engineer*, who is an active component of the operating system; and the *systems engineer*, working at the next higher level of abstraction in the systems hierarchy than that of the stationary engineer and dealing explicitly with what the stationary engineer treats intuitively. In management, the systems engineer 'discharges his responsibility by designing consistent and mutually enhancing relationships between social structure, information channels, and managing policy'. An ambitious attempt to realise this vision may be found in the Systems Engineering concept implemented at the Bell Telephone Laboratories. There, the systems engineers, senior people with an outstanding record of earlier scientific and technical achievement, also

[16] See Chapter 3 of this volume.

carry out the bulk of technological forecasting at both strategic and tactical level.

Planning may now be understood as the task of *designing a system* (a corporate policy), *flexibly changing subsystems* (corporate strategies) and the *ways to operate them* (operational plans). Technological forecasting furnishes much of the material to build the system.

FORECASTING METHODOLOGY

5. TOWARD A METHODOLOGY FOR SYSTEMIC FORECASTING
6. KEY APPROACHES TO FORECASTING

The development of a methodology for technological forecasting took a decisive turn, when a clear distinction was drawn between *exploratory* (opportunity-oriented) and *normative* (need-oriented) approaches. Iterating both approaches became a means for giving technological development a human and social perspective, and anchored forecasting firmly into the normative planning process. Of course, technological forecasting then becomes just an aspect of *integrative* or *systemic* forecasting. Technological, economic, or market forecasts are meaningful only at the operational planning level, i.e. for the short-range end of the planning continuum.

Chapter 5 gives a brief survey of methodological development struggling toward an improved capability for systemic forecasting. Most of this methodology is still in its infant stage. Chapter 6 sketches a few key approaches: Delphi, trend extrapolation, scenario-writing, morphological analysis, relevance trees, and cross-impact analysis. Both chapters propose various ways of classifying formal approaches, from different angles of view.

Chapter 5 is partly based on a presentation in the framework of the Policy Sciences Seminars of The RAND Corporation, Santa Monica, California, given on 9 December 1969. It was first published under the same title in *Technological Forecasting*, Vol. 1, No. 4 (Spring 1970), and is reprinted here by permission of American Elsevier Publishing Company, Inc., New York.

5. *Toward a Methodology for Systemic Forecasting*

INTRODUCTION

TECHNOLOGICAL forecasting in a narrow sense, aiming at forecasting technological feasibilities in terms of design or performance characteristics, belongs to a concept fitting a mechanistic model of innovative planning and action. Where the *purpose* of action comes into the focus of planning, the development and introduction of technology have to be viewed in a systemic context, in particular in the context of 'joint systems' of society and technology.[1] This is the case with full-scale, non-deterministic planning, as it unfolds in the interaction between the three levels stipulated by Ozbekhan[2]– normative (policy) planning, strategic planning, and operational (tactical) planning – and where the two upper levels, normative and strategic planning, bring the freedom of human choice into play, i.e. are conceived as aspects of a human action model. Normative and strategic planning are inherently 'open' (to inventiveness) and non-mechanistic, whereas 'closed' operational planning is employed for working out the strategies conceived, formulated, and assessed through the normative planning process.[3]

Forecasting has different tasks at these three levels.[4] At the upper two levels it forms an integral part of the planning process, its inventive core,[5] and thus involves notions such as purpose and norms. It focusses on outcomes, e.g., on systemic interactions of technology at the strategic level, and on the overall role of technology in structuring systems, in particular systems pertaining to both society and technology.

[1] See Chapter 3 of this volume.
[2] Hasan Ozbekhan, 'Toward a General Theory of Planning', in E. Jantsch (ed)., *Perspectives of Planning*, OECD, Paris 1969, p. 153.
[3] Hasan Ozbekhan, *op. cit.*, p. 118.
[4] See Chapter 4 of this volume.
[5] See Chapter 2 of this volume.

DIMENSIONAL INTEGRATION OF FORECASTING

Single-dimensional forecasting, e.g., technological forecasting in a narrow sense, can be meaningful only at the operational planning level where a clear-cut problem-solving approach is still valid for attaining well-defined technological targets, such as technical products or systems of a specified design and performance. Strategic forecasting, focussing on the purpose of introducing new technology, and on the systemic outcomes of its introduction, is inherently *integrative forecasting*,[6] a term which has come to denote a simultaneous approach – forecasting cutting across techno-logical, social, economic, political, psychological, anthropological, and other dimensions – combined with an active effort to achieve at least some sort of synopsis, if not an internally consistent or even truly systemic view.

We may here, in analogy to current discussions around pluri-, inter-, and trans-disciplinary approaches to education, attempt to distinguish between four levels of dimensional integration in forecasting:

—*Multi-dimensional forecasting* proceeds with parallel forecasts in various (isolated) dimensions, and views them synoptically in (not necessarily consistent) spatial cross-sections, without attempting spatial and temporal harmonisation; at this level, technological, economic, social, and other types of forecasting are defined in the traditional narrow sense.

—*Pluri-dimensional forecasting* proceeds first with parallel forecasts in various (isolated) dimensions, but attempts subsequently, for discrete time-depths, to set them in relation to each other and, through some spatial and temporal harmonisation, to obtain reasonably consistent spatial cross-sections; technological, economic, social, and other single-dimensional types of forecasting are meaningful here only as 'crude' inputs to pluri-dimensional forecasting.

—*Inter-dimensional forecasting* proceeds by establishing spatial and/or temporal causality relationships between discrete inputs pertaining to various dimensions, and simulates – by iteration or model simulation – spatial and temporal implications of their interaction.

—*Trans-dimensional forecasting*, finally, proceeds by establishing a spatial and temporal morphology of a complete system (however 'complete' may be tentatively defined here), from a specific normative angle of view, and aims at simulating total system behaviour; the normative view, coming to the foreground of interest today, is the ecosystemic view in the broadest sense, focussing on systemic (ecological) balances and imbalances.[7]

Single- and multi-dimensional forecasting presuppose, in particular, complete independence of technological development, and therefore are

[6] Frank P. Davidson, Macro-Engineering—A Capability in Search of a Method-ology, *Futures,* Vol. 1 (December 1968).
[7] Hasan Ozbekhan, *The Predicament of Mankind – Quest for Structured Responses to Growing World-wide Complexities and Uncertainties* (unpublished manuscript).

either largely unrealistic or – even much more dangerous – induce an attitude which accepts and reinforces an assumed 'autonomy' of technological development, and which finds its expression in Ozbekhan's warning formula 'The triumph of technology: "can" implies "ought".[8]

Pluri-, inter-, and trans-dimensional modes of forecasting recognise, to an increasing extent in this sequence, the systemic nature of occurrences in all dimensions, one of which is technology. It is evident that the focus shifts here from the development of technology to the development of societal systems, in particular the ecosystems of society and technology. This brings forecasting much closer to the forecasting of situations of a dynamic reality, and to elucidating both the complexity and the uncertainty inherent in reality. Forecasting of the inter- and trans-dimensional types is a prerequisite for enlightened choice, and generally for exercising the full normative planning process.

Progress from multi- and pluri- to inter-dimensional forecasting is characteristic of attempts to structure thinking and action by going to the next higher level of abstraction. For technology, for example, this implies thinking and action in a function-oriented framework, focussing on *functions* of technology in systems of society, such as communication, transportation, education, health, etc.,[9] with a variety of technological options grouping under each function. This already leads to organisational consequences in advanced-thinking industry.[10]

To a certain extent, inter- and trans-dimensional modes of forecasting need to be supported by individual contributions from single-dimensional forecasting, e.g., providing new ideas, estimates of technological potentials and limitations, and feasibility estimates.

'STATES OF MIND' IN FORECASTING

The degree of dimensional integration is not sufficient for characterising the applicability of forecasting approaches to the planning process. These approaches are not technical – or, as some reviewers prefer to say, scientific (a doubtful notion in planning, anyhow) – in the sense of being 'neutral'. Some of the formalised forecasting approaches correspond very closely to some of the basic modes of thinking quite generally (e.g., trend extrapolation, hierarchical structuring, morphological analysis). The formal approaches, or forecasting techniques, extend the capability of these basic modes of thinking by permitting one to handle more input information in a more homogeneous way. More than anything else, however, these

---

[8] Hasan Ozbekhan 'The Triumph of Technology: "Can" Implies "Ought"', in S. Anderson (ed.), *Planning for Diversity and Choice*, M.I.T. Press, Cambridge, Massachusetts 1968.

[9] See Chapter 8 of this volume.

[10] See Chapter 10 of this volume.

forecasting approaches represent 'states of mind', implying certain modes of use of these approaches.

Robert Theobald[11] has proposed a threefold classification of these 'states of mind': objective, subjective, and systemic approaches to forecasting. These are very useful distinctions, illuminating the aims of the forecaster:

— *'Objective' approaches to forecasting* are applied under the assumption that there is some objective, or near-objective truth to be found in the future. In its extreme form, the 'objective' approach corresponds to the notion of a 'futurology', as it may be encountered today in Germany, and to all sorts of naive futurism. But it also pertains to forecasting underlying operational planning and thus is applied both to linear planning (in current conventional planning, which is operational planning set absolute) and to the lowest level of the full-scale normative planning process, for the short-range, 'frozen-in' end only of the planning continuum, the rest of which is cybernetic.

— *'Subjective' approaches to forecasting* are applied under the assumption that the future can be shaped according to the preferences of some free-wheeling imagination. This is the other side of naive futurism. However, subjective approaches also dominate forecasting at the strategic level in the absence of a normative planning level, i.e., in the absence of criteria developed on the basis of total system dynamics and bringing value patterns and norms into play: this is generally the case with current governmental and industrial planning. Subjective approaches are, however, useful for strategic goal-setting in the framework of the full-scale normative planning process.

— *'Systemic' approaches to forecasting* are applied under the (valid) assumption that the systems of human living, or the world system in general, can develop toward discrete future states which depend on the spatial and temporal morphology of these systems. This implies that the directions in which these systems develop, depend on (forecasted), meaning invented changes in their morphology. We are, within certain boundaries, free to influence the structures of the systems in which we live, but the changes introduced determine the new boundaries within which we are free to plan and act. We are not free to aim at *any* imagined or preferred future.

It is evident that systemic approaches to forecasting, again viewed as 'states of mind' rather than technical tools, attempt to approach reality as closely as possible, whereas 'objective' and 'subjective' states of mind are characteristic of some form of escapism, or, *in extremis*, even of mysticism. It is equally evident, that only pluri-, inter-, and trans-dimensional forecasting modes correspond to a 'systemic' state of mind. The reverse, by the way, does not hold: Pluri- and inter-dimensional modes of forecasting, such as econometrics, may be applied in an 'objective' state of mind if the model is not flexible or 'open' enough to be manipulated; or a 'subjective'

[11] Robert Theobald, 'Alternative Methods of Predicting the Future', *The Futurist*, Vol. 3 (April 1969).

state of mind dominates where pluri-dimensionality appears in a hier-archical, non-systemic structure such as a relevance tree.

To a certain extent, formalised approaches which, in isolated use, correspond to 'objective' and 'subjective' states of mind, may be useful and even necessary for obtaining inputs to approaches characteristic of a 'systemic' state of mind. Examples are the tentative introduction of techno-logical feasibilities, contingent social or economic projections, value dynamics, etc., which may be based on single-, multi-, or pluri-dimensional forecasts. One may say that the simulation of potential structural elements and their attributes in a system precedes system simulation. The design of a dynamic system – or its model for simulation purposes – is an inventive task of high complexity, which depends on the availability of a wide range of 'building blocks', which all should be seen in a dynamic perspective.

DIRECTIONS OF FORECASTING

The distinction between exploratory and normative directions in forecast-ing serves as basis for the well-known fundamental classification of approaches to technological forecasting, proposed by Dennis Gabor:[12]

—The *exploratory direction of forecasting* leads from the present situation, along possible lines of development to future states. The temptation to explore along single-dimensional lines future states of discrete parameters is strong, the obligation to explore future situations, i.e., future states of a system in relation to its spatial and temporal morphology, difficult to fulfill. Exploratory forecasting, applied in isolation, tends on the one hand to reinforce linear planning, and on the other the indiscriminate pursuit of feasible activities.

—The *normative direction of forecasting* leads backward, from normatively assessed and judged future states to action in the present. If applied to planning which is restricted to the operational and strategic levels only, it may become reduced to *teleological forecasting*, a term proposed by Ayres.[13] If applied to the full-scale normative planning process, however, action oriented toward strategic goals and operational targets is ultimately invented, assessed, and selected in the light of criteria derived from norms; in other words, action is then judged as 'good' or 'bad', not just as being capable of reaching specified targets, or contributing to goals.

Quite generally, only an iterative or feedback process between explora-tory and normative forecasting provides useful inputs to the planning process, as is well known and recognised today. Nevertheless, unidirec-tional approaches to forecasting dominate the proliferous futuristic literature today, confusing issues and tasks of forecasting.

[12] Dennis Gabor, personal communication to the author, elaborated in, E. Jantsch, *Technological Forecasting in Perspective*, OECD, Paris 1967.

[13] Robert U. Ayres, *Technological Forecasting and Long-Range Planning*, McGraw-Hill, New York 1969, p. 33.

Normative forecasting leads to different basic approaches, if developed for a static framework on the one hand, and for a dynamic framework on the other. Underlying the static version is the conventional *sequential* problem-solving approach which suggests a hierarchical structure of inputs subsumed under outcomes (technological contributions to tasks and missions, R&D programmes contributing to technologies and products, etc.), as implied in the relevance tree approach. In contrast, feedback interaction in a dynamic framework suggests a systemic structure, in which hierarchical levels are not or not clearly and unambiguously distinguishable, and in which *consonance* with system dynamics, not hierarchical and priority ordering, is the key notion. Not even value patterns and sets of basic norms provide here, in the dynamic approach, unquestionable ordering principles from the top level of hierarchy, but undergo the same process of invention and 'try-out' in dynamic system simulation, as the other systemic elements and their attributes; only the system morphology is an expression of the values and norms brought into play. A formalised approach to forecasting, such as the relevance tree, therefore corresponds to a 'subjective', not 'systemic' state of mind, in spite of its linking vertically many domains, from social goals to R&D programmes.

Exploratory forecasting may correspond to a 'systemic' state of mind in both static and dynamic versions, depending on whether merely a spatial or a spatial/temporal morphology is aimed at. Normally, static versions cannot go beyond pluri-dimensional forecasting, with the interesting exception of time-independent contextual mapping which provides a multitude of spatial relationships without linking them explicitly to a time-frame.

STOCKTAKING: EXPLORATORY APPROACHES TO FORECASTING

Table 5.1 presents a survey of the main formalised approaches to forecasting – without any claim of completeness – arranged in a matrix by exploratory/normative directions and 'objective'/'subjective'/'systemic' states of mind. This tentative classification is based on present views on and uses of these approaches and should not be understood to express inherent limitations. As a matter of fact, the Delphi approach appears in three fields of the matrix, and multivariate models (econometrics) may come to be used in other than 'objective' states of mind. For each formalised approach, the degree of dimensional integration is indicated in *Table 5.1*.

A closer look at the exploratory approaches brings out clearly the evolutionary ladder which Roberts[14] finds in economic forecasting:

[14] Edward B. Roberts, 'Exploratory and Normative Technological Forecasting: A Critical Appraisal', *Technological Forecasting*, Vol. 1, No. 2 (Fall 1969).

States of Mind	Directions*	
	Exploratory	Normative

'Objective'	S-M Delphi
	S-M Trend extrapolation
	S-M Learning curves
	P Simple analytical
	  models and metaphors
	P 'Gap' analysis
	S-I Multivariate
	  (statistical) models
		S Operations research
		  (e.g., linear and dynamic
		  programming, resource
		  allocation models)
		S Network techniques

	I Normex	

'Subjective'	S SOON Charts
	S-M Delphi
	P Gaming
	P Heuristic programming
	  (statistical basis)
		S-M Delphi
		M-P IPIA methods
		  (psychoheuristic
		  programming)
		P Decision theory
		  (relevance tree,
		  decision models, etc.)

'Systemic'	(a) *static:*
	S-M-P Input/output
	S-M-P-I Morphological
	  analysis
	P-I Time-independent
	  contextual mapping
	(b) *dynamic:*
	P Scenario-writing
	P Cross-impact
	  analysis
	P-I Iterative system
	  projection
	P-I (Dynamic input/
	  output)
	I-T Structural models
		P-I Decision theory plus
		  systems effectiveness
		  concepts
		P-I Cross-impact analysis
		I Horizontal relevance
		  analysis
		T (Game theory)
		T ('Mendeleyev'
		  approach)

	T Learning models	

* The capital letters in front of each approach indicate the degree of dimensional integration: S, single-dimensional; M, multi-dimensional; P, pluri-dimensional; I, inter-dimensional; and T, trans-dimensional.

Table 5.1. A Survey of Forecasting Approaches, Classified by Directions of Forecasting and by 'States of Mind'.

1. Wisdom, expert, or genius forecasting (Delphi); 2. 'Naive' models (Trend extrapolation, simple analytical models and metaphors); 3. Simple correlative forecasting models ('Gap' analysis, gaming, heuristic programming, input/output, contextual mapping, scenario writing, iterative system projection, cross-impact analysis); 4. Complex multivariate econometric forecasts (Multivariate models, dynamic input/output); 5. Dynamic causally-oriented models (Structural models); and 6. Learning models. It also endorses, *grosso modo*, Roberts' assessment of technological forecasting as being essentially practised in approaches which do not go beyond step 3. Morphological analysis somehow falls outside this evolutionary sequence. If it can develop a dynamic version, it would correspond to step 5.

'Objective' exploratory approaches include the Delphi technique in its original application by Helmer and Gordon,[15] focussing on time-depths of future occurrences, a 'black box' use of the technique which brings into play a very heterogeneous basis of underlying assumptions which are not made explicit. The use of the Delphi technique for a 'subjective' approach may be considered much more appropriate and has, indeed, been promoted by users mainly in industry. Trend extrapolation is still the backbone of forecasting, with sophisticated versions, such as envelope curves,[16] only gradually finding practical use.[17] Learning curves[16] have probably not yet been explicitly applied to technological forecasting; their theoretical basis is not yet worked out. Simple analytical models and metaphors[16] (such as the biological growth analogy model) provide little more than a pseudo-rational explanation for the basic logistic curve encountered in trend analysis, correlating only few of the pertinent indicators; the potential for practical application appears as practically nil, and may remain so. 'Gap' analysis, as developed by Bouladon,[18] pointing out apparent opportunity ranges for technology, may become more widely used in combination with normative approaches. Multivariate models, finally, have not become known in tasks of technological planning, except for Floyd's[19] model for trend forecasting of figures-of-merit of complex technical systems. Econometric models have been refined so as to incorporate 'technology co-

[15] Theodore J. Gordon and Olaf Helmer, *Report on a Long-Range Forecasting Study*, Memorandum P–2982, The RAND Corporation, Santa Monica, California, September 1964; partly reprinted in Olaf Helmer, *Social Technology*, Basic Books, New York 1966.

[16] See, for example, the following monographs on technological forecasting: Robert U. Ayres, *Technological Forecasting and Long-Range Planning*, McGraw-Hill, New York 1969; and Erich Jantsch, *Technological Forecasting in Perspective*, OECD, Paris 1967.

[17] Marvin J. Cetron, *Technological Forecasting: A Practical Approach*, Gordon and Breach, New York 1969.

[18] Gabriel Bouladon, 'Transport', *Science Journal*, special issue on Forecasting the Future, Vol. 3 (October 1967); also: 'Technological Forecasting Applied to Transport', *Futures*, Vol. 2 (March 1970).

[19] A. L. Floyd, 'A Methodology for Trend-Forecasting of Figures of Merit', in J. Bright (ed.), *Technological Forecasting for Industry and Government*, Prentice-Hall, Englewood Cliffs, N.J. 1968.

efficients', expressing the effects on the economy of the diffusion of new technology.[20] It has already been pointed out above, that econometric models are not necessarily restricted to 'objective' uses, but are usually not applied in a 'systemic' state of mind because they are mainly based on statistical correlations, which do not bring out systemic structures and therefore the potential for purposeful manipulation of the model. An interesting combination of exploratory and normative 'objective' forecasting is the Normex approach developed by Blackman.[21] It has inter-dimensional potential in linking market characteristics and selected technological parameters. Normex forecasts are not merely concerned with 'pace-setting' technology, but also with the diffusion characteristics of technological innovation within specific market systems.

'Subjective' exploratory approaches include again the Delphi technique, but this time used for sharpening consensus on such factors as techno-economic feasibility and probability that an event will occur.[22] The ('objective' normative) network technique approach may be extended in an imaginative, 'subjective' exploratory, sense by seeking to recognise new opportunities created by realising step by step the course of action laid out by the normative network (e.g., new materials may be used for applications, not envisaged originally); this is the SOON Chart (Sequence of Opportunities and Negatives) approach.[22] If one of the values of Delphi is the 'involvement' of people at all levels, for example in an industrial corporation, it is perhaps the greatest benefit to be derived from gaming, and it combines there with the creation of a pluri-dimensional 'conscious-ness' of people in responsible positions.[23] Heuristic programming, an approach on which much emphasis is placed in the Soviet Union,[24] includes human behaviour on a statistical basis in the exploratory-stochastic model; it may be considered 'subjective' if it is conceived as a learning model, as is the case with the reported application, and 'objective' if the model is considered to be final.

'Systemic' exploratory approaches, finally, are of greatest interest here. As far as they aim only at some spatial morphology, they represent a static version, which, of course, is of limited use for planning under the influence

[20] E. Fontela, A. Gabus, and C. Velay, 'Forecasting Socio-Economic Change', *Science Journal*, Vol. 1 (September 1965).

[21] A. W. Blackman, Jr., 'Normex Forecasting of Jet Engine Characteristics', *Technological Forecasting and Social Change*, Vol. 2, No. 1 (Fall 1970).

[22] Harper Q. North and Donald L. Pyke, 'Technological Forecasting in Planning for Company Growth', *IEEE Spectrum*, Vol. 6 (January 1969); also ' "Probes" of the Technological Future', *Harvard Business Review*, May/June 1969.

[23] T. J. Gordon, S. Enzer, and R. Rochberg, 'An Experiment in Simulation Gaming for Social Policy Studies', *Technological Forecasting*, Vol. 1, No. 3 (Winter 1970).

[24] Alexander Y. Lerner, 'The Restoration of the Goal Function by Data on Prefer-able Responses' (paper presented at the International Seminar on Theory of Organisational Systems, Dubrovnik, Yugoslavia, 4–16 August 1969); published under the title 'A Learning Approach to the Dynamic Modeling of Human Planning and Decision-Making Systems' for *Technological Forecasting and Social Change*, Vol. 2, No. 2 (1970).

of complex dynamic systems. Input/output methods, so far used mainly for hindsight, are becoming increasingly applied to forecasting in research institutes,[20,25] but also at national level.[26,27] The bulk of applications so far has been single-dimensional, but some successful attempts to proceed to pluri-dimensional analysis may be recorded, even if all difficulties with quantification have not been resolved, as is usually the case with research/industry matrices, etc. The potential of input/output analysis for forecasting may be estimated to become much greater once a dynamic approach has been developed. Input/output representation of processes is under active development[28] and may ultimately be applied to the interaction between society and environment,[29] a very important idea in view of the emphasis being placed now on problems of this nature. Similarly, morphological analysis,[16] so far applied mainly in single-dimensional modes, may gain in importance through application to pluri- and inter-dimensional analysis of situations rather than technical configurations. Axiomatising 'systems of solutions' and 'systems of problems' and attempting to correlate them, as proposed by Leclercq[30] shows a way to the inter-dimensional application of morphological analysis. It may even become the approach *par excellence* to studying the alternatives inherent in complex societal situations, and it has an uncertain potential to be developed further into a dynamic approach. Time-independent contextual mapping, too, may be readily used in dynamic contexts, although it has deliberately eliminated temporal relationships in its approach; however, once a temporal relationship is tentatively introduced, the multitude of spatial configurations represented in the forecasts, instantly 'fall into place' also in the temporal morphology.

Dynamic approaches under the heading of 'systemic' exploratory forecasting include some relatively unambitious techniques, aiming not higher than at pluri-dimensional integration. This is the case with scenario-writing,[31] and its variation named iteration through synopsis;[32] however, scenario writing, as practised by Herman Kahn and his associates, is valuable for identifying forthcoming 'decision-nodes' and lessening 'carry-over' thinking. It is also the case with the version of cross-impact analysis,

[25] Erich Jantsch, *Technological Forecasting in Perspective*, OECD, Paris 1967, p. 208 f.

[26] Erich Jantsch, 'The Organization of Technological Forecasting in the Soviet Union', *Technological Forecasting*, Vol. 1, No. 1 (June 1969).

[27] N. P. Fedorenko, 'Planning of Production and Consumption in the U.S.S.R.', *Technological Forecasting*, Vol. 1, No. 1 (June 1969).

[28] Robert U. Ayres, personal communication.

[29] Anne Carter, 'Technological Forecasting and Input–Output Analysis', *Technological Forecasting*, Vol. 1, No. 4 (Spring 1970).

[30] R. Leclercq, 'Use of Generalized Logic in Forecasting', *Technological Forecasting and Social Change*, Vol. 2, No. 2 (Winter 1970).

[31] Herman Kahn and Anthony Wiener, *Toward the Year 2000: A Framework for Speculation*, Macmillan, New York 1967.

[32] Ronald Brech, *Planning Prosperity – A Synoptic Model for Growth*, Darton, Longman & Todd, London 1964.

developed at the Institute for the Future,[33] which is merely aiming at some harmonisation of probability estimates by establishing causal and pluri-dimensional links among small clusters of forecasted events, and adjusting forecasted probabilities in line with further ad-hoc assumptions. Iterative system projection takes essentially the same approach as implied by dynamic input/output forecasting, but reduces the difficulties greatly by keeping the system 'open' to some degree.[34]

As already recognised by Roberts,[14] the evolution of exploratory fore-casting techniques will inevitably lead to approaches based on the simula-tion of alternative systems configurations in structural inter-dimensional and, ultimately, trans-dimensional models. The most viable concept in this area is 'System Dynamics', originally known as 'Industrial Dynamics',[35] developed over the past decade by Jay Forrester and co-workers, which has also found applications to the dynamics of R&D.[36] Recently, it has been applied to the dynamics of cities[37] and of the world system.[38] It has the potential of being further developed to become a trans-dimensional approach to planning for complex social systems of concern to society.[39] 'Complex systems', in Forrester's terminology, refer to 'high-order, multiple-loop, non-linear, feedback structures', characteristic of all social systems. The present restriction of the approach due to internal inflexibility of the model during the simulation runs will probably lead to the develop-ment of heuristic and multilevel heuristic models.[40]

Learning models, combining 'systemic' exploratory and normative approaches, express the idea to link data bases of the present to data bases of the future (anticipations) in feedback models.[41] Whereas the exploratory direction may well be handled by the structural model approach outlined above, it is not yet clear how innovative changes in the present course of action should be derived automatically in the normative direction as a result of manipulations of the anticipations. Most likely, learning models

[33] T. J. Gordon and H. Hayward, 'Initial Experiments with the Cross-Impact Matrix Method of Forecasting', *Futures*, Vol. 1 (December 1968); also T. J. Gordon, 'Cross-Impact Matrices', *Futures*, Vol. 1 (December 1969).

[34] Aloys Schwietert and Wilhelm Nahr, *Westeuropa 1985/Western Europe 1985/ Europe Occidentale 1985*, Prognos A.G., Basel 1969.

[35] Jay W. Forrester, *Industrial Dynamics*, M.I.T. Press, Cambridge, Massachusetts 1961.

[36] Edward B. Roberts, *The Dynamics of Research and Development*, Harper & Row, New York 1964.

[37] Jay W. Forrester, *Urban Dynamics*, M.I.T. Press, Cambridge, Massachusetts 1969.

[38] Jay W. Forrester, *World Dynamics*, Wright-Allen Press, Cambridge, Massachu-setts 1971; see also Chapter 14 of this volume.

[39] Jay W. Forrester, 'Planning Under the Dynamic Influences of Complex Social Systems', in E. Jantsch (ed.), *Perspectives of Planning*, OECD, Paris 1969.

[40] M. D. Mesarović, in his conclusions for the International Seminar on Theory of Organisational Systems, Dubrovnik, Yugoslavia, 4–16 August 1969. See also Chapter 11 of this volume.

[41] Hasan Ozbekhan, *The Idea of a 'Look-Out' Institution*, System Development Corporation, Santa Monica, California, March 1965.

of a truly trans-dimensional character will become possible only through some man/computer interaction, where human inventiveness still accounts for the introduction of changes in the spatial and temporal morphologies which are then 'tested' by computer simulation.

It may be added in parentheses that the unsatisfactory state of probability concepts for forecasting may receive new impulses from the application of generalised logic (and the concept of plausibility)[30] which also has considerable potential for clarifying systemic interactions.

STOCKTAKING: NORMATIVE APPROACHES TO FORECASTING

The *'objective' normative approaches* are mainly single-dimensional. Operations research (from linear and dynamic programming to elaborate resource allocation models) maximises economic parameters from configurations of estimates, and network techniques for planning research and development link technical elements and requirements in what here may be justifiably called a teleological approach.

The *'subjective' normative approaches* are of great importance, because it is in the normative direction that subjectivity, in other words inventiveness and expressions of preference, have to be brought into play. Therefore, it also appears, that the use of Delphi as a 'subjective' normative technique exploits its best inherent potential. Such a use of Delphi focusses in industry on questions of 'desirability' (overall, and for the specific firm), and in societal contexts on values, as demonstrated in a fascinating experiment by the RAND Corporation.[42] IPIA methods (induced psychointellectual activity), which may also be called psychoheuristic programming, as described by Chavchanidze,[43] aim at giving research better goal-orientation by harmonising heuristic and subjective points of view. The most important formalised class of approaches in this category is based on decision theory, especially in the form of relevance trees, which may also underlie numerical approaches to obtain ranking orders, as exemplified by PATTERN.[16] These numerical approaches are particularly 'subjective' in so far as three types of estimates have to be introduced, and periodically updated, at each hierarchical level, namely criteria, weights of criteria, and significance figures. A similar rationale may be used to design decision models for computer operation. It has already been stated above that the hierarchical principle, which forms the basis of decision theory approaches, corresponds to a static view (even if technological development is introduced as being time-dependent and contributing to future goals). Attempts, to take

[42] Norman Dalkey and Bernice Brown, personal communication.
[43] Vladimir Chavchanidze, 'Induction of Psychointellectual Activity in Long-Term Forecasting and Planning of Research: Psychoheuristic Programming', *Technological Forecasting*, Vol. 1, No. 2 (Fall 1969).

at least cross-support of items at the same level (e.g., technological developments useful for more than one task) into account and thus to take a look at some 'R&D subsystem', have not led very far. More complex aspects of systemic forecasting, in particular feedback interactions, cannot be accommodated in this way.

The conceptually most important class of *'systemic' normative approaches* is not very rich in applicable techniques. A very interesting attempt to combine the 'subjective' decision theory approach with 'systemic' approaches at the individual levels of a relevance tree, has been reported by Fischer.[44] Subjective estimates of criteria, weights, and significance, are replaced here by more rigorous and more logical evaluations of systems effectiveness, which, however, may again base on estimates. Martin and Sharp[45] have developed an even more general approach to numerical relevance tree analysis by applying reverse factor analysis to criteria and significance figures, These approaches constitute significant attempts to go from pluri- to inter-dimensional normative forecasting. The same is attempted by those versions of cross-impact analysis which aim at linking estimates, pertaining to various dimensions, in a systemic way. If Šulc's approach to harmonise technological and social forecasts[46] remains pluri-dimensional, Pyke's[47] elaborate scheme for harmonising dimensional maps (e.g., environment, technology, applications maps), based on Delphi or other estimates, first internally, by means of normative networks and SOON Charts, and then in a sequence of steps in their cross-relations, already comes close to an inter-dimensional approach. The same holds, in principle, for horizontal relevance analysis[48] which is nothing else but an attempt to recognise systemic structures and make them explicit, without arriving at a complete systems model.

The only two ideas pertaining to trans-dimensional normative approaches, which would be badly needed to supplement dynamic systems simulation in the exploratory direction, remain theoretical, at least for the moment. The first one is the application of game theory, probably hopeless for the complex systems in question, but tempting with its promise of neatly calculated trans-dimensional optima. The second is the idea of a 'Mendeleyev' approach,[49] which would be capable of recognising the nature and form of interaction of systemic elements and their attributes, in

[44] Manfred Fischer, 'Toward a Mathematical Theory of Relevance Trees', *Technological Forecasting*, Vol. 1, No. 4 (Spring 1970).

[45] William Martin and James Sharp, 'Reverse Factor Analysis Approach to Relevance Trees', publication in *Technological Forecasting and Social Change,* Vol. 3 (1972).

[46] Oto Šulc, 'A Methodological Approach to the Integration of Technological and Social Forecasts', *Technological Forecasting*, Vol. 1, No. 1 (June 1969).

[47] Donald L. Pyke 'Mapping – A System Concept for Displaying Alternatives', *Technological Forecasting and Social Change,* Vol. 2, No. 3/4 (1971).

[48] William J. Sheppard, 'Relevance Analysis in Research Planning', *Technological Forecasting*, Vol. 1, No. 4 (Spring 1970).

[49] Hasan Ozbekhan, personal communication.

particular technologies before their identification (invention or specification), in an analogous way to the periodic table of chemical elements. The underlying idea here is that only discrete forms of interaction are possible in a system; one may therefore, to use another analogy, also speak of a quantum theory of societal or, generally, ecosystems. No theory or method has yet become known which has made much headway in this direction.

CONCLUSION

The growing recognition of the systemic nature of a world-wide *problématique* – 'the predicament of mankind'[7]– has contributed to the current emphasis on systemic approaches to forecasting and planning, which are also necessary ingredients to the full-scale normative planning process. The development of forecasting methodology has received significant impulses from this direction, as is borne out by the enrichment of approaches corresponding to a 'systemic' state of mind in forecasting, and by recent developments or extensions of older concepts in such approaches as structural models, horizontal relevance analysis, cross-impact analysis, input/output analysis, and iterative system projection, the combination of decision theory with concepts of systems effectiveness, heuristic and psychoheuristic programming, and network techniques. The aim is to proceed from pluri-dimensional approaches – to which the bulk of current 'system-minded' techniques belong – to inter- and further to trans-dimensional approaches. The only known forecasting approach with a clear (future) potential for trans-dimensional forecasting is the simulation of complex dynamic system models (structural models) on the exploratory side. The corresponding normative approach is still lacking; so are learning models, combining exploratory and normative forecasting in systemic feedback models.

6. *Key Approaches to Forecasting*

THE purpose of this chapter is not to give a comprehensive survey of forecasting techniques. The recent focus on technological forecasting has stimulated a wide variety of approaches which are described and set in perspective in the author's OECD survey[1] or Robert Ayres' book.[2] It is significant that most recent developments tend to refine some of the basic approaches known and applied already for many years, or even decades, rather than looking for new 'breakthroughs'. In particular, refinements are introduced to make forecasting more systemic, as outlined in Chapter 5.

It is a common misunderstanding that the use of techniques, or formalised approaches in general, should distinguish forecasting from mere speculation. Much of good forecasting is done without the explicit use of techniques. Techniques just serve to augment the capability of the forecaster and, in general, follow the basic thinking procedures which the human brain is applying intuitively. Most of them have been designed for a subtle 'man-technique' dialogue and are very sensitive to man's knowledge and his capacity for imaginative thinking, technical and value judgment, and synthesis.

The most important contributions of special techniques related to forecasting may be summarised in three points:

—They elucidate the role of individual input factors, compel a comprehensive consideration of such factors, and assure some homogeneity in the results.
—They tend to reduce prejudice and bias.
—They permit the evaluation of vast amounts and complicated patterns of input information and facilitate the systematic evaluation of alternatives.

If forecasting exercises are to relate in a meaningful way to corporate planning, they have to employ a variety of approaches and combine them in ways which depend on the nature of the forecasting task. Both exploratory and normative thinking, and thus the use of techniques belonging to

[1] Erich Jantsch, *Technological Forecasting in Perspective*, OECD, Paris 1967; about 100 variations of 20 principal approaches are identified and described in this survey.
[2] Robert U. Ayres, *Technological Forecasting and Long-Range Planning*, McGraw-Hill, New York 1969.

both 'directions', are necessary for a complete forecasting exercise. Simple techniques such as trend extrapolation, or scenario writing, may be used to generate information which is subsequently structured by other techniques and 'processed' for use in planning by still other procedures.

For the purpose of selecting the proper techniques, it might be of help to classify approaches to forecasting by their output – do they generate new information (which had not been explicitly on the table, although its elements might have been in people's minds), or do they simulate the use of this information? Again, the basic distinction between exploratory and normative approaches has to be made here. From this angle of view, forecasting techniques – explained here in terms of *technological forecasting* – may be generally grouped in the following way:[3]

1. *Techniques related to exploratory forecasting*, and all kinds of formalised approaches in the same direction, discover two important tasks: the generation of new information about future technological systems and their performance, and the simulation of the various outcomes of the realisation of (technological) options in the context of a diversity of possible situations.

   1.1. The *generation of new information* may be subdivided into extrapolative forecasts (where do trends lead to under linear or contingent assumptions?) and speculative forecasts (what is the range of alternatives?)

   1.1.1. *Extrapolative forecasting techniques* are mainly based on *trend extrapolation* and its refinements, of which the envelope curve technique is of particular interest (see discussion later in this chapter).

   1.1.2. *Speculative forecasting techniques* have achieved some sophistication in techniques for improving group consensus on intuitive thinking – ranging from brainstorming to the *Delphi* technique (see discussion below) – and in *morphological analysis*, systematically exploring all combinations of the qualitative variations the basic parameters of a concept (technological or other) can undergo, thereby bringing out the potential of new combinations (see discussion below).

   1.2. The *simulation of outcomes* of the realisation of options in various systemic contexts finds a variety of techniques, including *learning curves, gaming, input/output analysis, multivariate* and *structural models, scenario-writing* and *cross-impact analysis* (the latter two are discussed below).

[3] See also Donald L. Pyke, 'Technological Forecasting: A Framework for Consideration', *Futures*, Vol. 2, No. 4 (December 1970). In this paper, Pyke suggests a similar classification, and his distinction between extrapolative and speculative techniques has been adopted for this scheme.

2. *Techniques related to normative forecasting*, and formalised approaches in the same direction, also find two major tasks, again the generation of new information – but this time about needs, desires, values, functional requirements, and structural relationships – and the simulation of the implications of overall objectives (policies), goals (strategies) and specific operational targets in various systemic contexts.

2.1. The *generation of new information* may be subdivided into speculative techniques (what norms and what goals should we introduce into the planning process?) and structural techniques (what are the future relationships as affected by action we may take?)

2.1.1. *Speculative forecasting techniques*, in the normative direction, may again make use of improving group consensus through the Delphi technique (see discussion below).

2.1.2. *Structural forecasting techniques* have found their most elaborate example in *relevance trees* (see discussion below). Simpler applications of decision theory, such as decision matrices, are also in use, as well as network approaches to reasonably well-perceived goals. More recently, *cross-impact analysis* (which is also practised in a predominantly exploratory mood) has been developed as a means to structure and 'harmonise' future relationships in systemic contexts (see discussion below).

2.2. The *simulation of implications* of objectives and goals for action in the present again uses some of the structural approaches outlined above, such as relevance trees (in particular, in their numerical versions), all sorts of matrices or other simple procedures for priority ranking and rational resource allocation, usually basing on operations research and decision theory, dynamic modelling, occasionally game theory, and aspects of systems analysis. The aim of all these approaches is to guide the structuring of thinking by simulating the mutual consequences implied in the relationship between preconceived goals and recognised technologies or research elements. The criteria for investigations along these lines are generally preconceived, too.

A third class of techniques may be added here, which plays an auxiliary, if important, role in *'processing' forecasts* and linking them to corporate planning: at the level of operational planning most of these techniques may be grouped under the notion of economic analysis, based on cost/profit evaluations, in particular versions employing discounted cash flow approaches (and thereby taking the time factor correctly into account); at the strategic level, little has yet become known of more sophisticated cost/effectiveness approaches (except in the military domain) which would be of paramount importance for the assessment of forecasted alternative options.

The *Delphi* technique for the improvement of intuitive thinking through the sharpening of group consensus, was developed mainly by Olaf Helmer[4] and his associates at the RAND Corporation. It is essentially a refinement of the original 'brainstorming' technique. The difference is that in seeking the views of many experts, procedures are used to make these experts sharpen their own thinking and to prevent them from exchanging opinions with one another. Delphi panel members never meet, but communicate with a 'control centre' (never with other panel members), usually by a succession of 'rounds' of letters. In most Delphi exercises, panellists are asked in the first round to enter suitable topics (questions). In an exploratory 'objective' application of Delphi, for example, experts may be asked what break-throughs they expect in a certain domain (e.g. information technology, or biomedicine, or space exploration) within the next 50 years. In the follow-ing rounds, estimates of various kinds will be asked of the experts, typically relating to the timing of specific events (e.g. the breakthroughs in the above example), to probability, feasibility, desirability and so forth. After every round, answers received from the panellists are synthesised at the 'control centre', and information about the range and distribution of estimates is fed back to the panellists. The result of time estimates, for example, will be communicated to them in form of indications for the median and the quartiles of the total group estimate: If half of the panellists estimate the event to happen before 1987, and the other half after that date, 1987 is the median value; if the most 'optimistic' quarter of the answers lies before 1982, and the most 'pessimistic' quarter of the answers after 1996, the lower and upper quartiles for the specific event or breakthrough are 1982 and 1996, respectively. (In industry, indications of where the extreme lower and upper ten per cent of the answers lie are sometimes considered more appropriate.) The principal aim of the Delphi exercise is now to narrow down this spectrum of answers in iterative rounds, by giving the panellists a feeling of where their estimates are located within the group synthesis and also by asking them for reasons for possible extreme positions of their answers. After three or four rounds of estimates on the same question, the above example of a time estimate may have changed to a median of 1988, and lower and upper quartiles of 1986 and 1992. Group consensus has considerably sharpened, and confidence increased that the event will take place in the six-year span between 1986 and 1992, and more likely toward the nearer end of it.

Estimates of probability, feasibility, and desirability may be quantified in any way. In some good industrial Delphi exercises a scale from -1 to $+1$ is chosen; zero would then be the indifference point, whereas negative

[4] Olaf Helmer, *Social Technology*, Basic Books, New York 1966.

figures would express degrees of improbability, unfeasibility, and un-desirability.

As indicated in *Table 5.1*, the Delphi technique may be applied in an 'exploratory' as well as a normative direction. However, using it for explo-ratory 'objective' forecasting, in other words, for prediction – for example, estimating the time and probability of scientific breakthroughs, or social change – implies the conscious or unconscious use of different sets of assumptions by the experts, and is of limited value. Only 'point' forecasts of more or less isolated events can be predicted in this way – but prediction is a meaningful notion only at the level of operational planning, as has been pointed out in the preceding chapters.

In more sophisticated uses in the framework of corporate planning,[5] the 'subjective' state of mind dominates. The Delphi technique may find its most important applications in fields where intuitive thinking will remain one of the principal sources for inputs and consensus is important, as in the formulation of goals at all levels. In industry, a company-wide Delphi exercise has, in numerous cases, been found an excellent means to 'involve' a large number of people and also to stimulate future-oriented thinking beyond the initial exercise. In a Delphi exercise, each (anonymous) panellist's estimate carries equal weight – the Chairman's as well as a junior engineer's. No individual opinion gets 'lost' in the process and none dominates. Usually, panellists are assured of the strict anonymity of specific estimates, especially in industrial Delphi panels.

Trend extrapolation is the oldest forecasting technique and has long played a role in economic forecasting and other single-dimensional cuts through corporate dynamics. The extrapolation of trends over time usually assumes that the specific constellation of forces influencing growth over the recent past, will continue its influence in the future, possibly with some marginal modification. Thus, trend extrapolation is the technique *par excellence* for an 'objective' state of mind in forecasting, and has become most effective in reinforcing linear planning and the 'bureaucratic' aberration sketched in *Figure 2.4*. If used wisely, however – which means, for example, in the perspective of a multi-contingency approach (assuming alternative constellations of influencing forces, such as level of effort, funding, people, incentives for research, etc.) – it is one of the principal generators of new knowledge which may subsequently be fed into more elaborate and more systemic forecasting approaches. Applied to techno-logical forecasting, a number of refinements has been developed, of which three will be briefly discussed in the following: S-curves, extrapolation by precursory events, and envelope curves.

S-curves, or *logistic curves*, are the graphic expression of growth under restraint, in particular under the influence of absolute or practical limits,

[5] Harper Q. North and Donald L. Pyke, ' "Probes" of the Technological Future', *Harvard Business Review*, May/June 1969.

not many of which are usually known in advance (which makes trend extrapolation of technological growth risky). Instead of an exponential growth – if the growth rate remains constant in dependence of the level achieved – the curve goes through an inflection point, after which growth diminishes. This implies also that there is a diminished return on further efforts, in terms of technological capability – but often also in terms of commercial or other utility (*see Figure 6.1*).

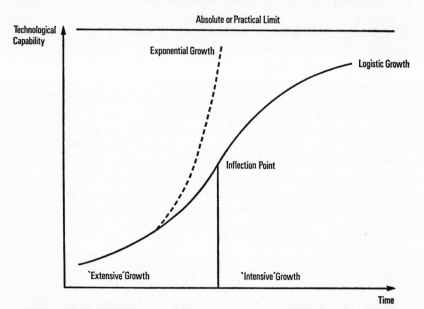

Fig 6.1   Trend Extrapolation.   Not exponential, but logistic growth (S-curve) is a realistic assumption. It is important to know the level of saturation which inhibits free growth – often, it can be determined only by fundamental research programmes directed to this question.

Extrapolation by *precursory events* recognises that a particular development trend 'leads' another development trend, especially where basic technologies may be shared. The most striking example is the growth of transport aircraft speed, which is 'led' by combat aircraft speed (*see Figure 6.2*). It should be noted, however, that strong normative components coming into technological goal-setting now may well break the smooth curves and also interfere with the close linkage between the two development curves.

Envelope curves, finally, represent thinking which is refined and risky at the same time. Few forecasters in industry dare to employ it systematically so far. The basic idea is that a broad technological capability – for example, speed attained by man (*see Figure 6.3*) – may be extrapolated in the same way in which capabilities of a specific technology may be

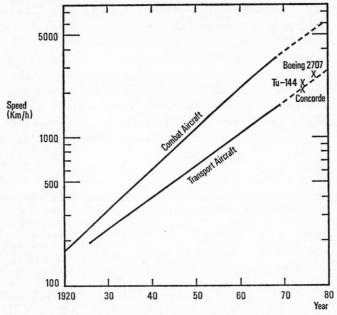

Fig 6.2   Trend extrapolation by precursory events

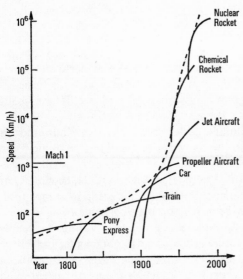

Fig 6.3 Envelope curve extra-polation is an important means of forecasting technological capability. The maximum speed attained by man increases in conformity with an S-shaped envelope curve; individual technologies tend to level off at their respective 'saturation' level. Forecasting by extrapolating the envelope curve essentially means forecasting the impact of a forthcoming breakthrough, without defining the technology which will achieve it.

extrapolated. But that technological capability is now seen as being not restricted by the maturing of a specific technology, but subject to being pushed further by a succession of different technologies. The result is usually a 'big S' (the envelope curve) riding on a series of 'small S's'. From a purely exploratory angle of view, this opens up the fascinating prospect of 'looking around the corner' and forecasting future states of technological capabilities without any idea about which technology will make their attainment possible. If a forecaster feels that he can place high confidence in his trend analyses, he may risk a decision to take up a new technological option at a time when this new option is neither as good as the best

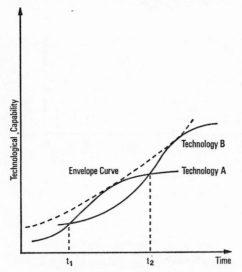

Fig 6.4 Advantages and pitfalls of envelope curve extrapolation are illustrated in this graph. A company exploiting technology A can gain considerable advantage by recognising and developing technology B before A starts to level off. On the other hand, direct comparison of trends A and B at time t_1 or shortly afterwards would lead to the wrong long-range conclusions if not seen in the context of the envelope graph. Few firms can see situation t_2 at time t_1.

available (mature) technologies, nor improving as fast as they do (*see Figure 6.4*). This was the situation when the Xerox technology was offered to a number of big industrial corporations, which all turned it down on grounds of the discouraging superficial evidence (it was finally taken up and developed by the Battelle Memorial Institute).

Scenario-writing, the technique perfected by Herman Kahn and his associates at the Hudson Institute, is essentially an attempt to view and combine various trends in a systemic way. Typically, small single-dimensional steps are taken (for example, in technological, economic, or political developments), the changes implied for the scenario are mapped, and decisions which have to be made (or may be made) due to these changes are then systematically explored. In this way, various sequences of small logical steps into the future are explored to obtain a feeling for the long-range systemic consequences. The primary purpose is not to predict the future, but systematically to explore branching points dependent upon critical choices. In fact, a vast or almost infinite multitude of possible scenarios can be generated in this way, dependent on the decisions tested in the process. It becomes clear, that it is meaningless to speak of one scenario for the year 2000. Scenario writing is more serious than that – it is a formalised approach to mapping alternative courses of action into the future.

Morphological analysis, in the words of its pioneer Fritz Zwicky who applied it as early as 1942, is 'an orderly way of looking at things' and so achieving 'a systematic perspective over all possible solutions to a given large-scale problem'.[6] This is an exploratory approach which attempts to break up the problem into its basic parameters, and then conceive of as many variations of each parameter as possible. In this way, for instance, Zwicky has shown that a simple chemical jet engine characterised by 11 parameters (for each of which two, three, or four variations can be conceived) can be thought of in 36,864 possible combinations of these 11 parameters – and after the systematic removal of internal inconsistencies, there remain 25,344 consistent design possibilities. How Zwicky wants his method to be used to break bias in the thought process – or apply bias only at the very end of the analysis – is demonstrated by the variations he introduces for the parameter 'surrounding medium': vacuum, air, water, earth. Jet engines operating in an air environment are commonplace today. Hydro-jets operating in water seem, at first glance, not impossible. But terra-jets digging themselves through earth? The point is that such possibilities should be discarded only after the analysis has been made, if at all, for being of no practical use or too far in the future, etc. This final bias should not affect the morphological analysis itself. *Table 6.1* shows an imaginative hypothetical example in the textiles field. Another example of morphological analysis, for the food production function, is given in *Table 8.3*. The morphological approach can be applied to all sorts of situations, from fundamental research over technology all the way to social, political, and economic problem areas, and it may also become very useful for systemic approaches.

Relevance trees, or vertical relevance analysis, constitute another 'orderly way of looking at things', but from the other end, starting with broad objectives and goals and attempting to set up hierarchical relationships for all conceivable contributions to them. Considerable guidance for planning may already result from such a structuring of potential activities, from fundamental research all the way over technological systems to the missions they may be able to serve and to concepts of corporate or societal roles in general. Relevance trees constitute, in fact, the only systematic approach in use linking the entire range of levels. This, however,

[6] Fritz Zwicky, *Morphology of Propulsive Power*, Monographs on Morphological Research No. 1, Society for Morphological Research, Pasadena, California 1962; Zwicky's approach is described in detail in Erich Jantsch, *Technological Forecasting in Perspective*, OECD, Paris 1967.

		1	2	3	4	5	6	7	
A	Effect	Colour	Brightness	Handle	Water repellency	Oil repellency	Pesticidal	→	etc.
B	Reagent	Dye (pigment)	Oba	Softener	Wax	$F^{c/ds}$	Mitin	→	etc.
C	Medium	Water	Solvent	Vapour	Air	Emulsion	Substrate	→	etc.
D	Process	Absorption	Reaction	Precipitation	Diffusion	'Dry-on'	Adhesion	→	etc.
E	Substrate	Cotton	Wool	Nylon	Polyester	PAC	Paper	→	etc.

Examples for known technologies:

A1 – B1 – C1 – D1 – E1: Direct Dye Cotton
A6 – B6 – C1 – D1 – E2: Mitin Mothproof Wool
A1 – B1 – C1 – D4 – E5: Maxilon Dye Agrylics

Examples for feasible new technologies:

A2 – B2 – C1 – D2 – E6: Reactive Tinopal Paper
A6 – B6 – C1 – D2 – E2: Reactive Mitin Wool
A1 – B1 – C6 – D2 – E3: Reactive Pigment Nylon

Table 6.1. Morphological Analysis, applied to the textiles field.[7]

[7] Gordon Wills, 'The Preparation and Deployment of Technological Forecasts', *Long Range Planning*, Vol. 2 (1969).

implies that fairly precise ideas must be available at each level, e.g. what technological systems may be perceived to contribute to certain tasks, what subsystems they would require, what technology deficiencies would have to find a solution, and what R&D programmes might contribute to these solutions. This makes relevance trees fully applicable to technological forecasting only within a time-frame of 10–15 years, in most cases. This limitation will be even more stringent for numeric versions of the relevance tree, of which Honeywell's PATTERN (Planning Assistance Through Technical Evaluation of Relevance Numbers)[8] has become most famous. *Figure 6.5* gives an example of a PATTERN application (only the qualitative relationships show in the figure). Criteria, numerical weights for them, and 'significance numbers' (e.g. how significant a technological system's potential contribution to a specific task is) are usually estimated and periodically updated by expert panels. The aim is to rank possible activities, e.g. potential R&D programmes. Fischer[9] has recently developed an approach to measure relevance in a more objective way by describing utility at a specific level through a multi-parameter space and looking at the advances a particular contribution promises in this multi-dimensional space (the advances, though, will have to be forecast). There are many potential pitfalls in using the relevance tree approach, in particular in its numerical version (where errors in the higher levels affect strongly the relevance numbers obtained for the lower levels, e.g. for R&D programmes. It is often not possible to enter a good estimate of the importance of future objectives, goals, and missions – because a relevance tree typically is supposed to represent relationships that will hold 10 years hence. Therefore, every construction of a relevance tree starts with writing plausible scenarios to get a feeling for these future relationships. Also, usually, a large number of individual forecasts (e.g. trend extrapolations) are made in order to fill in future technological options, future states of basic technologies (e.g. materials), etc. Morphological analysis may be used to generate alternatives at the higher, more complex levels. Toward the end of the 1960s, relevance trees have come into wide use for industrial forecasting, in America as well as in Europe. In most cases, qualitative versions only are applied.

Horizontal relevance analysis also has a great potential for structuring possible activities and their impacts in a systemic context. In fact, horizontal relevance analysis is nothing else but plotting a number of important systemic relationships of a non-hierarchical character. It is a useful,

[8] Maurice E. Esch, 'Planning Assistance Through Technical Evaluation of Relevance Numbers', *Proceedings of the 17th National Aerospace Electronics Conference*, Institute of Electrical and Electronics Engineers, New York 1965, pp. 346–51; this technique is also described in detail in Erich Jantsch, *Technological Forecasting in Perspective*, OECD, Paris 1967.

[9] Manfred Fischer, 'Toward a Mathematical Theory of Relevance Trees', *Technological Forecasting*, Vol. 1, No. 4 (Spring 1970).

though first-approximation, approach to building parts of system models. It may be carried further – although it is hardly practised in this way today – in expressing the spatial configuration of complex feedback relationships and intermeshing feedback loops. This leads to structural models of the 'System Dynamics' type which may serve as the basis for computer simulation. However, the simple first step of plotting relevant relationships between factors pertaining to many dimensions (techno-logical, economic, social, etc.), is already another useful tool for structuring inputs to planning.

Cross-impact analysis is the attempt to recognise causal links or chains between 'point' forecasts (events), usually arrived at through Delphi exercises, and to modify scenarios accordingly to make them more internally consistent. Two different versions have been proposed so far. One has been developed by the Institute for the Future in Middletown, Connecticut,[10] and has been imaginatively applied to various problem areas. In the preceding Delphi exercises, events are forecast together with estimates of timing and of probability that the event will take place. Then, causal links are identified and both enhancing and suppressive modes of interaction are considered. Hypothetical examples may be easily con-ceived: An event A (a social fashion, such as a preference for a particular performance of an urban transportation system) may give rise to specific efforts and funding and thereby enhance the probability for event B (the maturing of a specific technology, such as the linear induction motor) to happen within a certain time period. Event B, in turn, may trigger a higher level of scientific research in a particular area and thereby enhance the probability of event C (a scientific breakthrough, such as super-conductivity at environment or at least easily controllable temperatures). To calculate the new probabilities, usually a quadratic function of the time differential is (arbitrarily) assumed for the enhancing and suppressing modes.

Donald Pyke[11] proposed a more sophisticated approach to cross-impact analysis, which, however, has so far not been tested in practical use (it would require very considerable effort). 'Point' forecasts (events) are divided into three broad categories – environment, applications, and basic technologies – and further subdivided into narrower categories (e.g. pro-duct lines within applications, scientific or technological areas for basic technologies). For each of the three maps – 'Enviromap', 'Applimap', and 'Technomap' – causal networks are built of the events, and new events (of

---

[10] T. J. Gordon and H. Hayward, 'Initial Experiments with the Cross-Impact Matrix Method of Forecasting', *Futures*, Vol. 1, No. 2 (December 1968).

[11] Donald L. Pyke, *The Role of Probe in TRW's Planning*, paper presented at the conference Technological Forecasting: An Academic Inquiry, Department of Manage-ment, Graduate School of Business, The University of Texas, Austin, April 1969. See also by the same author: 'Mapping – A System Concept for Displaying Alternatives', *Technological Forecasting and Social Change*, Vol. 2, No. 3/4 (1971).

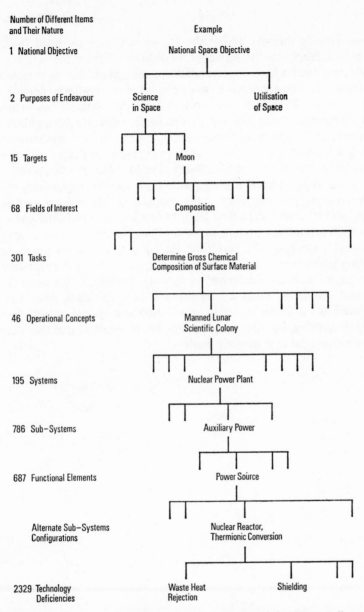

Number of Different Items and Their Nature	Example

1 National Objective — National Space Objective

2 Purposes of Endeavour — Science in Space / Utilisation of Space

15 Targets — Moon

68 Fields of Interest — Composition

301 Tasks — Determine Gross Chemical Composition of Surface Material

46 Operational Concepts — Manned Lunar Scientific Colony

195 Systems — Nuclear Power Plant

786 Sub–Systems — Auxiliary Power

687 Functional Elements — Power Source

Alternate Sub–Systems Configurations — Nuclear Reactor, Thermionic Conversion

2329 Technology Deficiencies — Waste Heat Rejection / Shielding

Fig 6.5 PATTERN, a relevance tree scheme developed by Honeywell, was applied by NASA to its Apollo programme payload evaluation.[12]) On the lower levels of the relevance tree 'commonalities' occur so frequently that, for example, 301 different tasks can be achieved by means of only 46 operational concepts. Criteria and numerical weights (not shown here) are introduced at each level so that 'relevance figures' can be calculated for the research and development programmes needed to resolve the technological deficiencies; in this way priorities can be established for the allocation of funds. This is an example of normative forecasting: the goals are known but means to achieve them must be specified.

[12] W. S. Beller, 'Technique Ranks Space Objectives', *Missiles and Rockets*, 7 February 1966.

which nobody thought during the Delphi exercise) are filled in where necessary. New possible networks, branching off from the original networks, are built where opportunities are recognised (e.g. new materials, developed for one application, may open up opportunities to develop new applications) – Pyke calls them SOON Charts (Sequence of Opportunities and Negatives). Networks within the three maps are harmonised and brought into logical order. Subsequently, in a number of steps, harmonisation is attempted between the three maps in respect of timing, estimates of probability, feasibility and desirability. For example, the desirability of a new technology which is a prerequisite for a specific application, cannot be lower than the desirability of the application. This sequence of 'inter-map harmonisation' steps then leads to considerations of total corporate potentials and limitations (e.g. in terms of investment, manpower, research potential, marketing and servicing, etc.) and finally to an input/output representation for the entire company which may, in an ultimate step, be harmonised with an anticipated input/output table for the national, or regional economy. What distinguishes Pyke's approach from the IFF approach is the emphasis on harmonisation and spatial as well as temporal structuring, i.e. 'moulding' the original forecasts so that they become more meaningful in a systemic context.

THE FOCUS ON SOCIETAL FUNCTIONS OF TECHNOLOGY

7. GUIDING BASIC RESEARCH
8. FUNCTION-ORIENTED PLANNING OF TECHNOLOGY

Normative thinking can give spur and guidance to all phases of the technological innovation process – including research at fairly fundamental levels. Such guidance becomes effective through a framework focussing on *outcomes* instead of (material and non-material) inputs, in particular on the functions technology is expected to perform, or contribute to, in the systems of human living – functions such as communication, transportation, health, education, power generation and distribution, food production and distribution, etc., along with their subfunctions. The shift from product-, technology-, skill-, or resources-oriented thinking to function-oriented thinking in industry implies breaking out of 'technological imperatives' and gaining a new flexibility in bringing alternative technological options into play. It marks the expansion from mere operational to *strategic* planning and action. Examples are elaborated for the chemical industry and for the food production function.

Chapter 7 was first published in French under the title 'Prévision Technologique et Recherche Fondamentale' in *Atomes*, No. 246, September 1967. It appears here for the first time in the original English version by permission of La Recherche/Atomes, Paris.

The bulk of Chapter 8 is extracted from two previously published papers: The first three sections are adapted from a presentation at the First Technological Forecasting Conference, organised by the Industrial Management Center (Prof. James R. Bright) in Lake Placid, New York, in May 1967; they were first published under the title 'Integrating Forecasting and Planning Through a Function-Oriented Approach' in James R. Bright (ed.), *Technological Forecasting for Industry and Government; Methods and Applications*, 1968, and are reprinted here by permission of Prentice-Hall, Inc., Englewood Cliffs, N.J. The section 'A Function-Oriented Framework for the Chemical Industry' is adapted from a presentation at the Tripartite Chemical Engineering Conference, Montreal, Canada, 22–25 September 1968; it was first published under the title 'The Transformation of the Chemical Industry and Consequences for Chemical Engineering' in *The Chemical Engineer*, No. 229, June 1969, and is reprinted here by permission of The Institution of Chemical Engineers, London. The last section, 'A Technological Decision-Agenda for the Food Production Function', has not been previously published.

7. Guiding Basic Research

INTRODUCTION

'IT is the choice of problem more than anything else that marks the man of genius in the scientific world', Sir Henry Tizard stated in the 1930s. However, fundamental research is still frequently mistaken to stand, by definition, for undirected ('pure') research, and the two terms are widely used as synonyms. For about 50 years, science and technology have grown together in the technological innovation process and fundamental research, being at the origin of all major technological innovation, generally has to be at least broadly oriented if it is supposed to trigger or contribute to desired innovations. Effective coupling between fundamental research and the applied research and development phases, a most important and at the same time most delicate problem which also seems to be at the roots of the much-discussed 'technological gap' between America and Europe, can be improved by technological forecasting which attempts to focus the direction of the technological innovation process. Technological forecasting, on its way to becoming a serious activity, and finding widespread use in industry since the early 1960s, can become an effective means to direct fundamental research to areas pertinent to industrial, national, or social goals. At the same time, fundamental research enters into a close dialogue with technological forecasting providing the answers to questions on basic potentials and limitations, thereby greatly improving the forecasting function. Both the nature and the volume of fundamental research will be increasingly determined by this dialogue, in the industrial as well as in the public domain.

The Committee on Science and Public Policy (COSPUP) of the U.S. National Academy of Sciences – National Research Council initiated, in 1964, the large-scale systematic assessment of the social value of research in the fundamental scientific disciplines, and its potential contribution to broad social and national goals.

An American aerospace company already reaps some benefits of its own research in molecular biology (eliminating bacteria in fuel tanks) and investigates the fundamental requirements for superconductivity at non-cryogenic temperatures and the potentials of meeting them with new organic substances. The Xerox Corporation is embarking on a large-scale fundamental research programme on photoconduction in a wide spectrum of materials in order to assess the alternatives to current reproduction technology. The information and communications sectors develop new lines of mathematical logic and take an interest in brain neurology, with the ultimate aim to transgress the present boundaries of syntactic communication (where only the structure is considered) to semantic (structure plus meaning) and pragmatic (structure plus meaning plus effect on recipient) types of communication.

In the early 1950s, a considerable improvement in the thermo-electric 'figure of merit' was achieved through the use of semiconductors, then a new class of materials; Westinghouse went straight ahead with a fairly large-scale development programme for thermo-electric household refrigeration, whereas General Motors (Frigidaire) directed first fundamental research to the same problem and recognised correctly a natural limit for semiconductors which was generally still below the economic threshold. In American defence research, the attained temperature resistance of materials in an oxydising atmosphere – a functional capability whose improvement was pushed with all available means because of its paramount importance for missile and other defence developments – tended to level off far below the assessed natural limit; a hitherto unknown 'disguised' intermediate limit made itself felt. A major effort, involving top experts, was undertaken by the Institute for Defense Analyses and succeeded in determining the fundamental nature of that intermediate limit (related to the oxygen transport mechanism in the material); the required breakthrough was subsequently achieved.

These are just a few examples, representing different orders of importance and of research volume involved, to demonstrate that fundamental research can be effectively directed. One may supplement them by examples in which the lack of direction in fundamental research has proved an obstacle for innovation which conceivably could have sprung forth from it:

Penicillin, for ten years after its genuinely accidental discovery by Fleming in 1929, was not adapted for medical uses. Whereas metals and semiconductors, in the focus of interest through clearly formulated technological needs, have benefited enormously from fundamental solid-state research, the ceramics sector has been left in a shamefully empirical state due to the lack of recognised incentives, which, however, exist. Funds in the order of 100 million dollars per year, which the U.S. Department of Defense used to place with universities for undirected research, have essentially failed to produce the expected effect of triggering a burst of technological innovation, 'Project Hindsight'[1] revealed.

[1] Raymond S. Isenson, et al., *Project Hindsight*, First Interim Report, October 1966. Clearinghouse for Federal Scientific and Technical Information, Springfield, Virginia 22151 (U.S.A.).

THE CHANGED RELATIONSHIP BETWEEN SCIENCE AND TECHNOLOGY

It is significant that a now obsolete – but still widely repeated – notion attempted to define fundamental research by the motives, or rather non-motives, for it being undertaken: it was supposed to serve the advancement of scientific knowledge without 'specific commercial objectives' (or 'specific products', or even 'any practical aim at all') in mind. While this formulation in psychological terms had always to be regarded as arbitrary and artificial, it nevertheless fitted most of the fundamental research until approximately 50 years ago when the present strong coupling between fundamental research and technics started to develop.

A. P. Speiser[2] recently pointed out this revolutionary change: 'Until approximately 50 years ago technology and science developed pretty independent from each other. The steam engine was invented, before even the elementary principles of thermodynamics had been formulated; the first telegraph was built, before the so-called telegraph equation had been discovered, i.e. the equation that describes the transmission of an electrical signal along a wire. . . . In other words: Technology would have progressed practically in the same way if scientific research (then performed almost exclusively at the universities) would not have existed or would have been entirely different. . . .' The example of aircraft may also be mentioned here which became a technical success before the correct theory of lift and drag had been formulated; on the basis of the available wrong calculations the famous astronomer Simon Newcomb, in 1903, called flight 'one of the great class of problems with which man can never cope' – only eight weeks later the first flight was achieved by the Wright brothers.

'Since then,' Speiser continues, 'this relationship has changed drastically. The first technical breakthrough which came from fundamental science, is wireless communication. . . . Maxwell formulated the electromagnetic theory and predicted waves in the ether; Hertz designed an experiment which confirmed the existence of such waves; Marconi repeated Hertz' experiment and pushed development to a point where large-scale technical and industrial exploitation became possible. . . .'

'Today all revolutionary technological innovation has its roots in fundamental research. Nylon, the transistor, and nuclear energy are impressive examples for this thesis. It has become almost impossible that a gifted inventor succeeds with a revolutionary innovation by making use only of known physical principles.'

[2] Ambrosius P. Speiser, 'Was erwarten wir von der naturwissenschaftlichen Forschung?' and 'Die Forschung in der Schweiz', in *Neue Zürcher Zeitung*, issues of 1 January and 26 January 1967 (foreign edition), Zurich (Switzerland).

A meaningful notion of fundamental research today is devoid of any motivational content (which may still be used to signify the degree to which it is directed). Fundamental research is simply research on the fundamentals of science and technology. It precedes or interacts mainly with the 'discovery' and the 'creation' phases of technology transfer, as represented in *Figure 7.1*. The farther technology transfer proceeds in its course towards technological innovation, the more focussed become research, development and engineering. Only the general movement can be seen as a time-sequence, whereas individual action often takes the reverse direction, for example when fundamental research becomes necessary to cope with newly encountered problems. However, the degree to which research is 'pure' and to which it is directed becomes less and less meaningful in science policy, and represents only a secondary characteristic derived from the inherent laws of technology transfer between the more general and the more specific. One might argue whether the mapping of radiation belts around the Earth or the search for a theory of super-conductivity in absolute terms (i.e. going beyond the phenomenological BCS theory) represents 'pure' research or not. It can be taken as significant that industry favouring technological innovation increasingly aligns its thinking and planning to the time-sequential RD&E phases as indicated in *Figure 7.1*, whereas the distinction between the classical RD&E categories becomes more and more blurred, due to the width of the research spectrum contributing to the early phases of technology transfer (represented in the figure by the thickness of the arrow near its origin).[3]

There is obviously a deeper sociological reason for the persisting confusion between fundamental and undirected ('pure') research, primarily in the realm of science policy in the public domain: the wish of many scientists, whose work is supported by the public, to maintain the worn-out notion of 'academic freedom' by claiming that fundamental research defies any attempt of planning. 'Relinquishing the intellectual throne for the life of a commoner is a hard chore for science after three hundred years of free ranging and a hundred years of lordship.'[4]

Whereas science has been flexibly adapted to its new role in interaction with economic and military developments, much of the science policy in the public domain is still based on a view of science that is now obsolete by 50 years. The changes which are gradually becoming visible in modern concepts of science policy, so far mainly in the United States, will be the more brutally felt.

---

[3] Compare also *Figure 4.1* for a forecasting and planning approach in terms of the RD&E phases.
[4] R. H. G. Siu, *The Tao of Science*, M.I.T. Press, Cambridge, Massachusetts 1957.

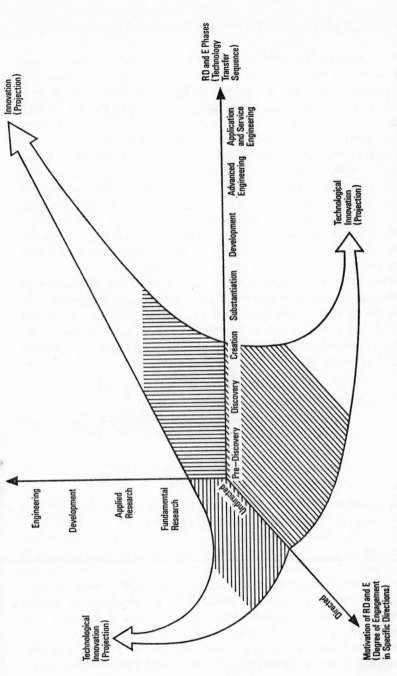

Fig 7.1 The role of fundamental research in technological innovation, graphically viewed in terms of projections of a three-dimensional representation. The individual RD&E Phases (research, development and engineering phases as defined by the Stanford Research Institute) correspond to the degree to which research is directed. The domain of fundamental research is represented by the shaded areas, penetrating to a considerable extent into directed research.

THE COUPLING FUNCTION BETWEEN SCIENCE AND TECHNOLOGY

Since fundamental research has become a prerogative for revolutionary technological innovation, the coupling between fundamental research and the later RD&E phases, as one moves along the technology transfer sequence, becomes of crucial importance. It has been stated on both sides of the Atlantic, that basic scientific knowledge and skill in Europe generally competes well with American achievements – but it is also true that it now leads to less technological innovation. It is only gradually becoming recognised now that the coupling function is weak in Europe, and that one of the primary reasons for any 'technological gap' between the two continents can be found there – and in Europe's lagging behind in management technology which can be applied to improve the coupling. As a matter of fact, a 'technological gap' does not exist in a drastic form where Europe has understood to use her fundamental science more effectively as, for example, in some of the chemical sectors.

Many examples of weak coupling can be studied in the history of technical developments. A group of technological innovations, which have emerged very slowly from fundamental research, and have not yet fully attained their target, is provided by the technology of direct conversion from heat to electricity. The scientific principles were discovered back in the 19th century:

1822 Seebeck demonstrated the thermo-electric effect;
1834 Peltier demonstrated the reverse effect;
1838 Faraday measured the magnetohydrodynamic effect induced by the flow of the dirty river Thames (carrying ions) in the magnetic field of the earth;
At the end of the 19th century Edison discovered and Thomson investigated thermionic emission.

No technological developments sprang from these discoveries until a second step in fundamental research, motivated by purely scientific incentives, had been achieved in the 1920s: Both thermo-electric and thermionic effects could be understood (with discrepancies remaining) only after modern theoretical physics by means of wave mechanics, the Pauli exclusion principle and other theoretical refinements resulting in the concept of Fermi statistics had elucidated the phenomena of electric conduction in solids. Ionisation phenomena in gases and plasma physics in general had to await theoretical interpretation and accumulation of evidence even from astrophysical observations, to prepare magneto-hydrodynamics and to some extent also thermionics for practical exploitation. It was not until the 1950s when, due to the strong incentives from space and defence programmes, fundamental and applied research became much better focussed and an upsurge of directed research programmes pushed technology transfer further to a point where specific applications came within reach, and materials problems became predominant and may urge more research to support development and engineering. Similarly, the technology of superconductivity had to wait for incentives to be pushed on the basis of scientific and technological elements that would have been available since the 1920s.

However, in the words of Harvey Brooks, 'history is a very poor guide; we have improved'. American industry traditionally not only has considerable faith in science – and commits large funds to it – but is also well aware of the long-range potentials of in-house fundamental research. General Electric founded its research laboratory, part of which has always been active in fundamental research, already in 1900 – long before the above-mentioned basic shift in original 'creativity' from technology to science began to dawn on European electrotechnical industry. Today, the creation of 'Science Centres' for independent (company-funded) research and development, with much emphasis on fundamental research, is spreading even among the big government contractors – for example, in the aerospace sector. In Europe, where interaction between universities (and their fundamental research) and industry has been much less pronounced so far, some of the advanced-thinking industrial groups are now in the process of changing to a policy attaching more importance to the performance of in-house, and the utilisation of external, fundamental research.

In the public domain, the 'percentage principle' which largely reflects the influence of pressure groups and still dominates the allocation of funds for fundamental research has essentially failed to produce the desired results. Whereas the idea to maintain a certain balance in the advancement of science is not without merit, the hopes that new lines of technological innovation will be initiated in this way were not fulfilled.

On the basis of abundant evidence from technological innovation today there is strong support for the hypothesis that effective coupling in technology transfer can only be achieved in the presence of normative guidance. In other words, the broad requirements for technological innovation have to be given if fundamental research is supposed to trigger a technology transfer 'chain reaction'. Einstein, who himself was the recipient of an unequalled series of 'flashes of genius' in 1905, pertaining to three different areas of fundamental physics, supported expressly a highly pragmatic view: 'The history of scientific and technical discovery teaches us that the human race is poor in independent thinking and creative imagination. Even when the external and scientific requirements for the birth of an idea have long been there, it generally needs an external stimulus to make it actually happen; man has, so to speak, to stumble right up against the thing before the idea comes.'

The lack of clearly formulated requirements – which obviously exist – has so far prevented the success of molecular biology in the fundamental area to penetrate considerably into the applied areas. The recognised requirement of a rational pharmacology may be an exception and is guiding part of the research in this fertile field. Holography, invented in 1948 in Great Britain on theoretical grounds by Dennis Gabor and becoming feasible with the later developed laser light sources, has been successfully picked up for technical developments in America to contribute to a number of recognised missions – although its original inventor was among the first to stimulate normative thinking in Europe by his appeal to 'invent the future'.

Guidance to fundamental research may become effective already when it is applied in very broad terms only. The 'show window' examples of revolutionary technological innovation, the transistor and nylon, have both succeeded because (a) the 'fertility' of a field of fundamental knowledge – solid-state physics and macromolecular chemistry, respectively – had been recognised, and (b) a clear overall objective was formulated. In the case of the transistor, the presence of this overall objective (a decisive innovation in communication technology, leading to the pursuit of the alternatives semiconduction and electroluminescence) was so imperative that the direction of research was changed almost instantly when – instead of getting closer to a 'suspected' field effect amplifier principle – the first hint to another amplifier principle, the point contact transistor, was found; fundamental research, prompted thereby, led to the theoretical prediction and subsequently the realisation of the junction transistor, three years later. The repeated reformulation and refinement of the normative guidelines evidently added much impetus to the technology transfer from the level of fundamental science to the finished product, and is now systematically applied to all work at the Bell Telephone Laboratories.

In the equally well-known examples of nuclear energy and radar development, the readily available elements of fundamental science and technology were put together and directed towards these innovations only when the strong normative stimulus of the threatening war became effective.

TECHNOLOGICAL FORECASTING – A COUPLING DEVICE

'Der Herrgott ist raffiniert, aber boshaft ist Er nicht' (The Lord is subtle, but he isn't simply mean) – the truth of Einstein's statement is of basic importance for the performance of fundamental research. It means, as Norbert Wiener[5] has emphasised with much weight, that the level of

[5] Norbert Wiener, *The Human Use of Human Beings*, Houghton Mifflin, Boston 1950; Second Revised Edition, Anchor Book A 34, Doubleday, Garden City, New York 1954.

fundamental research is favoured by one condition which holds for no other level traversed in the course of a technology transfer process: The environment of fundamental science and technology does not 'react' to man's research, an objective can be pursued by a strategy which does not have to take counter-strategies by Nature into account. Here, and only here, the time factor is not implicit but is introduced by man alone. Forecasting is reduced to the recognition of invariable patterns of objectives, criteria, and relationships, and to the assessment of man's capacity to approach them, and of the pace at which this can be done.

In spite of this state of affairs favourable to the inclusion of the funda-

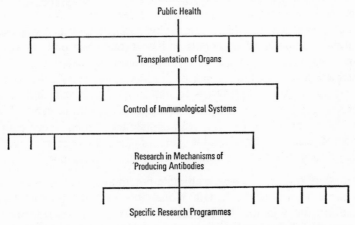

Fig 7.2 Application of the relevance tree approach to problems of fundamental research, as proposed by J. Bronowski. Numerical versions of the relevance tree technique may also be applied.

mental level in technological forecasting, much less attention has thus far been devoted to this area than it deserves. There is no doubt that the 'purist' attitude of scientists has acted as an inhibition to treading on their territory.

The incentive to forecasting at the fundamental levels is also enormous from another point of view: any error, or misdirection, committed at these levels is reflected in large and expensive failures. Recognition of this incentive has, for example, inspired the American military to adopt a policy stressing technological forecasting at the fundamental levels.

Figure 7.2 illustrates in a simple way how the relevance tree technique might be applied to guide fundamental research to attain a recognised social goal. This approach is now used more and more frequently to translate broad future goals into structural terms. In an exploratory direction, morphological analysis is finding interesting applications in attempts to bring the potential of fundamental research into play in a systematic way. Morphological analysis has been applied successfully to astronomy, for example.

Technological forecasting can interact with fundamental science in three different ways:

1. It can, on the basis of available knowledge on fundamental potentials and limitations, attempt to forecast conceivable results of future fundamental research (for example, the feasibility of utilising subnuclear energy or annihilation energy); this is the least important task, however.
2. It can aid to direct fundamental research to areas which are pertinent to broad objectives and high-level goals; this task is in the foreground today in industry and is rapidly gaining attention in the public domain.
3. Technological forecasting and fundamental research can enter into an intimate dialogue in which, essentially, forecasting puts the questions (ultimate or practical potentials and limitations, the bulk of which is unknown) and fundamental research provides the answers; this fruitful interaction which, at the same time, improves technological forecasting considerably and contributes to the stimulation and focussing of research, will revolutionise part of the fundamental research, first in industry – already a few American companies with advanced planning concepts align a considerable part of their fundamental research accordingly, thereby changing its nature and increasing its volume.

An interesting way to assess ultimate potentials has been proposed by Sir George Thomson, the British Nobel Prize-winning physicist:[6] their derivation from principles of impotence, i.e. laws that are statements of the principles of Nature that cannot be violated (for example, the laws of conservation of mass and energy say that any interchange between the two follows a strict mathematical relationship). He advances the thesis that the truly basic laws of physical science are such principles of impotence.

How far the direction of fundamental research in a 'streamlined' science-based company can go may be illustrated by the example of one of the leading semiconductor companies which has established an 'innovation emphasis structure' throughout the company which resembles a relevance tree, whose two bottom levels are marked 'strategies' and 'tactical action programmes'. Whereas approximately 90 per cent of the company activities in the operating divisions tie in with this structure, still as much as 75 per cent of the corporate-level central laboratory activities, representing a large share of fundamental research, contribute explicitly to the tactical action programmes and thereby to the supreme company objectives.

The systematic and comprehensive approach which technological forecasting represents, will also stimulate the recognition of inter-

---

[6] Sir George Thomson, *The Foreseeable Future*, Cambridge Univ. Press, Cambridge (England) 1955, revised edition 1960.

disciplinary potentials as well as 'commonalities' in the research along different lines of technological innovation. Apart from other effective means, such as mobility of people between the fundamental and the applied research areas, and the use of consultants by industry, technological forecasting has already become one of the most important coupling devices between fundamental science and technology.

FUNDAMENTAL RESEARCH AND SOCIETY

The old controversy whether science supported by the public should be directed or not is gradually being lost by the defenders of the 'purist' attitude. It is interesting to note, as Taton has pointed out,[7] that some of the great French mathematicians of the 19th century already tried to relate the value of their science to the broad goals of society. Fourier's belief that the chief aim of mathematics is its public usefulness and its explanation of natural phenomena led to a controversy among mathematicians at that time. Poincaré did not pursue a subject because of his mental resourcefulness but because of the needs of science. In giving a triple aim for mathematics – physical, mathematical, and aesthetic – he insisted that the physical and aesthetic aims 'are inseparable and (that) the best means of obtaining the one is to look at the other or at least never to lose sight of it.'

Today the recognition that in our 'Technical Age' the advances and the application of science and technology have become the most powerful means for transforming society, gradually forces public science policy to direct fundamental research more and more to areas which are pertinent to national and broad social goals. Once we understand that we are, within wide limits, free to choose our own fate by guiding technological development, the recognition and evaluation of alternative futures (de Jouvenel's 'futuribles') becomes the most important task. Technological forecasting provides effective tools for translating our future into structural terms right down to the level of fundamental science, for example, by applying relevance tree techniques.

A first important qualitative approach to this end can be seen in the systematic evaluation of fundamental science on the basis of Alvin Weinberg's[8] 'internal criteria' (ripeness of a field, availability of good research people, etc.) and 'external criteria' (scientific merit, including impact on related scientific fields, technological merit, social merit) by the Committee on Science and Public Policy (COSPUP) of the U.S. National

---

[7] R. Taton, *Reason and Chance in Scientific Discovery*, translated from French, Science Editions, New York 1962.
[8] Alvin M. Weinberg, 'Criteria for Scientific Choice', *Minerva*, 1963, pp. 159–71.

Academy of Sciences – National Research Council. A number of topics investigated illustrate the role of technological forecasting in this exercise:[9]

—Present and future status of the field;

—The future programme: assignment of priorities, recommendations;

—Substantive questions and unanswered questions;

—Modes of attack and levels of understanding;

—New tools;

—Opportunities presented;

—Influence on concepts in other scientific areas;

—Influence on techniques in other scientific areas;

—Impact on technology, applications, etc.;

—Relation to the industrial economy and to defence; opportunities and problems for industry and science;

—Manpower requirements for the next five years;

—Manpower forecasts for five- and ten-years time-depth.

An attempt to assess a field of science which is rapidly moving from the fundamental to the applied levels, namely oceanography, on the basis of a cost/benefit analysis, failed because it used an erroneous mathematical basis.[10] In general, however, a straight cost/benefit analysis seems to be applicable only to a small fraction of fundamental research in the public domain. Quantification on a cost/effectiveness basis, which has been successfully applied to military planning, offers much greater potential; it can attempt to assess national and social objectives and weigh the potential contribution of fundamental research. An important beginning has been made by the introduction of the medium-range Planning-Programming-Budgeting System (PPBS) to American civilian government on the basis of systems analysis and cost/effectiveness studies. Further developments in mathematical language to link costs with values, and to bridge the entire range from high-level goals (end-uses) to fundamental research, i.e. the full technology transfer process in social areas, seem possible.

[9] U.S. National Academy of Sciences – National Research Council, *Physics: Survey and Outlook – Reports on the Subfields of Physics*, Washington, D.C., 1966.

[10] U.S. National Academy of Sciences – National Research Council, Committee on Oceanography, *Economic Benefits from Oceanographic Research*, Publication 1228 (out of print), Washington, D.C., 1964.

8. Function-Oriented Planning of Technology

THE CASE FOR STRATEGIC THINKING IN INDUSTRY

TRADITIONALLY, forecasting has been regarded as distinct from planning. It can even be practised with the conscious or unconscious purpose of evading planning, if a deterministic view of the future is taken. 'Planning will force action', as a high European government official remarked sadly to the author, whereas purely exploratory forecasting is sometimes accepted as depicting the inevitable future – one's own future included – so that there is little left to influence or change through planning.

Human achievement has never sprung from such a defeatist view. The ancient Greek tragedy – the strongest expression man has given to his notion of historical forces that are beyond his power of control – emphasises that, of his own free will, man has to accept the verdict of history and that his fate will be fulfilled only if he aligns his own forces with those of the gods. This subtle communication of man with history, which looks like a basic paradox only in its superficial aspects, has underlain the dynamic development of Western civilisation from the time of the Greeks to the present. On the verge of a more thorough and rapid transformation of society than the world has ever seen – the transformation by 'social technology' – man is again called upon to take decisions of his own free will, to make this breathtaking turn of history actually happen.

In spite of the dramatic technological successes of the first half of this century, a surprisingly passive attitude toward technological development has dominated until recently – an attitude which was content with primarily exploratory (opportunity-oriented) forecasting. This is exemplified by comments concerning television which were made in 1936 (only one year before regular television broadcasting started), by two of the most courageous pioneers of technological forecasting in America. The metallurgist C. C. Furnas, in a collection of remarkable technological forecasts,[1] wrote:

[1] C. C. Furnas, *The Next Hundred Years – The Unfinished Business of Science*, Williams & Wilkins, Baltimore, Md., 1936.

'I am waiting for my television but I cannot live forever. When I think that the first radio impulse transmission was accomplished by Joseph Henry in 1840 and the first radio broadcast was not until 1920, I feel a little discouraged about the arrival of this television business while my eyes still function. No one has dared even to think of television in natural colours as yet.'

And S. Colum Gilfillan, who had pioneered technological forecasting in the United States since 1907, asked at the same time: 'Will the public accept TV and pay for it?'[2]

Even after the successful demonstration of 'crash programmes' such as those which led to the development of nuclear energy and radar in World War II – the result of primarily normative (mission-oriented) thinking – the impact on industry was not felt immediately. A 'Forum on the Future of Industrial Research',[3] held in New York in 1944, with an impressive array of some of the most distinguished names from U.S. Government, industry, and universities, arrived at a consensus favouring a concept based on such notions as: no planning of research; encouragement of co-operative research; and only small science-based industry which would not 'plan', but would 'try out' new products. This concept for the future, which essentially expressed a desire to return to pre-war habits, changed drastically only when the incentives were again stimulated by the Cold War. Europe has not yet fully caught up with this switch.

A new attitude toward the future is developing today. Based on increased confidence, gained and already 'tested' in military technological development, it now penetrates from philosophical levels to applications in socially significant areas, and it is starting to transform industrial thinking. Olaf Helmer, when he planned his 'Institute for the Future' – which is now operating in Middletown, Connecticut – best describes this new attitude toward the future:[4]

'The fatalistic view that it is unforeseeable and inevitable is being abandoned. It is being recognised that there are a multitude of possible futures and that appropriate intervention can make a difference in their probabilities. This raises the exploration of the future, and the search for ways to influence its direction, to activities of great social responsibility. This responsibility is not just an academic one, and to discharge it more than perfunctorily we must cease to be mere spectators in our own ongoing history, and participate with determination in moulding the future. It will take wisdom, courage, and sensitivity to human values, to shape a better world.'

Technological forecasting will contribute to an inter-disciplinary, integrative analysis along these lines, and the main difficulty of making

[2] S. Colum Gilfillan, 'The Prediction of Inventions', in U.S. Natural Resources Committee, *Technological Trends and National Policy*, Washington, D.C., 1937.

[3] Standard Oil Development Company, *The Future of Industrial Research*, New York 1945.

[4] Olaf Helmer, *et al.*, *Prospectus for an Institute for the Future*, Santa Monica, California 1966.

such thinking fruitful will have to be faced: the conscious or unconscious tendency to select future goals along lines of thought that are valid only for the present. Technological forecasting makes it possible to break out of the extended present – the 'logical future' – and to penetrate into a 'willed future' (René Dubos). It not only links with technological forward planning today, but is well on its way to becoming an integral part of futures planning in the broadest sense and one of the principal tools for 'futures-creative' thinking (Hasan Ozbekhan).

The main features of modern technological forecasting, which has been finding wide applications in industry since about 1960, differ radically from the older notions that went with speculative and, generally, purely exploratory forecasting. These features have been discussed in some of the previous chapters of this book. Technological forecasting, gradually taking shape along these lines to become a valuable management tool, has already entered a relationship of intimate interaction with planning. In many companies today, it is still considered a separate activity and is applied in discrete time-intervals to focus and spur planning. The two can be expected to enter complete symbiosis in the 1970s, when industrial thinking will be increasingly affected by the basic new orientations that are already becoming visible:

1. The shift of emphasis from short- and medium-range tactical to long-range strategic thinking; and
2. The trend toward large-systems thinking, going beyond the industrial and economic systems to include more and more of the future societal systems.

These important implications were grasped already more than ten years ago by Jay W. Forrester in his visionary 'Industrial Dynamics' concept, which is perhaps only now starting to bear real fruit:

'Historically, military necessity has often led not only to new devices like aircraft and digital computers but also to new organisational forms and to a new understanding of social forces. These developments have then been adapted to civilian usage. . . . As the pace of warfare has quickened, there has of necessity been a shift of emphasis from the tactical decision (moment-by-moment direction of the battle) to strategic planning (preparing for possible eventualities, establishing policy, and determining in advance how tactical decisions *will* be made). . . . Likewise in business: as the pace of technological change quickens, corporate management, even at the lower levels, must focus more and more on the *strategic* problems of running the business and less and less on the everyday operating problems. In the system development for the military, it has been amply demonstrated that carefully selected formal rules can lead to tactical decisions that excel those made by human judgment under the pressure of time and with insufficient experience and practice. Furthermore, it has been found that men are just as adaptable to the more abstract strategic planning as they are to tactical decision making, once their outlook has been lifted to the broader and longer range picture.'[5]

[5] Jay W. Forrester, 'Industrial Dynamics – A Major Breakthrough for Decision Makers', *Harvard Business Review*, Vol. 36, No. 4 (July–August 1958), pp. 37–66. The

With corporate managements of some of the innovation-minded American companies already devoting more than half of their time to a future ten or more years distant, the truth of this statement is finding dramatic proof today.

THREE DYNAMIC CONCEPTS OF INDUSTRY

Long-range strategic and large-systems thinking has already transformed the organisational schemes of the United States and other governments. The *Planning-Programming-Budgeting System* (PPBS), in use in the U.S. Department of Defense since 1961, and introduced into the civilian branches of the U.S. Government in 1965, is, on the outside, a medium-range planning scheme with five- to six-year time-depth. However, it encourages, and even enforces, long-range strategic thinking beyond the formal planning scope, by virtue of the following particular features: the upgrading of medium-range forecasting and planning to a continuous activity which shapes thinking and decision making in the entire government area and directs government agencies to *functions* relevant to national objectives and broad societal goals; a systems-oriented approach which is quite new in government, including the evaluation of alternatives; and the use of advanced techniques such as systems analysis, operations research, cost/effectiveness studies, and model-building.

With fully operational PPBS, which is still a vision that is only gradually materialising in the civilian U.S. Government, research and development forecasting and planning (i.e. the 'technological innovation structure' of the government) will be similar to that of a decentralised company, stressing decentralised initiative and centralised balancing. The departments and agencies will develop their 'businesses' along the lines of the decentralised operating divisions of a company and will compete for funding in accordance with their contribution to 'corporate objectives'. The Executive Office of the President can be compared, in this framework, to corporate-level staff groups. (Such a 'business model' of government would be utterly resented in most of the European countries today.)

The most important aspect of this system is the abolition of the traditional service structure in the military and the instrumental structure in civilian government, and their replacement by a functional framework for forecasting and planning. Weapon systems development is, in the 'corporate' picture, no longer considered in the framework of the armed services – Army, Navy, and Air Force – but, instead, is seen in functional programme categories such as: 'Strategic Retaliatory Forces', 'Continental Air and Missile Defense Forces', 'Civil Defense', 'Airlift and Sealift',

concept, which aims at finding a useful basis for model-building to apply to complex dynamic systems, was subsequently expanded in Jay W. Forrester, *Industrial Dynamics*, M.I.T. Press, Cambridge, Massachusetts 1961.

etc. In 1967, the Canadian armed forces went even a step further in abolishing their service structure altogether, even for administrative purposes. In the civilian branches of the U.S. Government, similar structures of functional programme categories are identified and gradually adopted for planning.

Where does *industry* stand in this search for new organisational forms to incorporate strategic and large-systems thinking? Today, a number of underlying factors are undergoing subtle but thorough changes, and tend to form characteristic patterns. The rough classification according to *Table 8.1* may be understood as a progressive sequence of patterns related to industrial dynamic concepts. Certain industrial sectors (in their broad layout, not in their organisational structure) have always been in line with function-oriented thinking; e.g., from the beginning, all the big electro-technical groups, both in Europe and America, have recognised their power engineering function to be an integrated power-generation/power-distribution/power-utilisation function (combining vastly different product lines in thermal and electrical machinery and apparatus). The functions that tie in with the product-line class 'electronics' have been less clearly defined. Not all of the big electrotechnical groups have entered the data processing field, and some have left it after an unsuccessful attempt.

Figure 8.1 illustrates three typical forecasting attitudes toward a simple technological capability such as speed, strength, etc. Step 1 – 'Drifting with the main-stream' – is characteristic of what may be called the 'bandwagon' attitude. It can provide a comfortable basis for existence, but will inevitably lead to frustration of the type the bigger European countries are now labelling the 'technological gap'. Also, more sophisticated analysis will probably reveal that the reputation and strategic position of a company are becoming increasingly dependent on the innovation it has to offer, not so much on the high quality of its conventional output. Whereas in conventional power plant building, the turbogenerator manufacturer played the key role and usually subcontracted the boiler, etc., the client for a nuclear power plant now turns to the manufacturer of the nuclear 'boiler' first, with the turbogenerator assigned relatively secondary importance. As a consequence, all the big electrotechnical groups – none of which had been active in the conventional boiler field – have entered, or are trying to enter, nuclear reactor development. It may be expected that the step-1-attitude will have even more difficulty competing with a future step-3-environment than with the currently predominating step-2-attitude.

Today, most companies fit into the pattern corresponding to step 2 – 'leading the main-stream' – which has dominated the American industrial scene for the past couple of decades, and is now winning over Europe. (Of course, not everybody who tries will become a leader; it is the attitude, not the result, which is discussed here.) At this step, normative thinking already

Dynamic Concept	1. 'Drifting with the Main Stream'	2. 'Leading the Main Stream'	3. 'Shaping the Future'
Principal incentive	Hedging against threat; maintenance of average position	Maintenance of leading position (entrepreneurship, restricted to departmentalised thinking)	Challenge (entrepreneurship, development of competence, etc.)
Emphasis in future-orientation	Short-range tactics (up to 5 years)	Medium-range tactics (5 to 10 years)	Long-range strategy (10 years and beyond)
Principal optimisation criterion	Low-risk profit	Maximum profit	Corporate objectives tying in with societal objectives and goals
Typical organisational structure	Horizontal (mainly centralised initiative and control)	Vertical (decentralised initiative in distinct product lines, centralised control)	Function-oriented innovation emphasis structure, penetrating throughout the entire company and encouraging entrepreneurship at all levels; 'matrix management' or basic split between 'Present' and 'Future' at high levels
Forecasting emphasis	Mainly exploratory	Strong normative component gives spur and guidance to product-line planning	Evaluation of strategic large-systems options in feedback cycle with planning
Planning emphasis	Mainly indicative, if at all	Operational planning, focussing on product lines	Flexible strategic planning, in functional framework, encouraging 'competition between alternatives', systemic context

	Forecasting in a corporate-level staff group or committee	Decentralised forecasting and planning in product lines, confronted with each other at discrete time intervals, at top mainly financial control	Integrated, continuous forecasting and planning at all levels, co-ordinated at the top
Typical organisation of forecasting and planning			
Estimated year of 'management innovation' (considerable impact)	Traditional	U.S.: 1953/54 Europe: 1963/64	U.S.: 1965/66 Europe: 1969/70
Estimated share in big industry with a high rate of technological innovation (for 1970)	U.S.: Few Europe: 50 per cent	U.S.: ~90 per cent Europe: 50 per cent	U.S.: ~10 per cent Europe: Few

Table 8.1. The Progressive Trend in Dynamic Concepts of Industry.

HUNT LIBRARY
CARNEGIE-MELLON UNIVERSITY

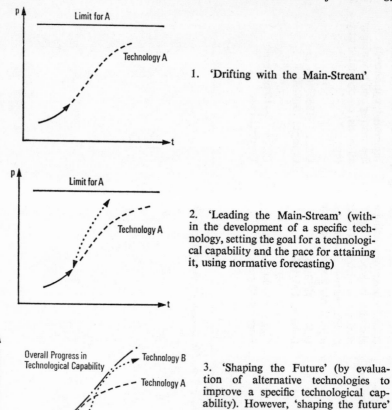

1. 'Drifting with the Main-Stream'

2. 'Leading the Main-Stream' (within the development of a specific technology, setting the goal for a technological capability and the pace for attaining it, using normative forecasting)

3. 'Shaping the Future' (by evaluation of alternative technologies to improve a specific technological capability). However, 'shaping the future' may also mean the pursuit of a different technological capability contributing to a technological and societal function.

Fig 8.1   Typical Forecasting Attitudes (p = technological capability, t = time)

goes far beyond the clear-cut problem of defining requirements for a specific mission and participates in formulating the missions themselves. This approach is representative of the current notion of 'free enterprise'. However, it is basically restricted by the piecemeal approach it offers. Its inadequacies can be demonstrated today in many ways and at all levels. Consider the following examples:

—Subsidies and 'government guarantees' for continued coal mining are common in a period when coal is rapidly losing ground in the inter-fuel competition;

—The utilisation of fossil hydrocarbons for combustion purposes is still expanding, although it is now becoming clear that the natural reserves could and should be used as valuable raw materials for upgraded products, such as food and textiles.[6]

Product-oriented thinking prevails, not only in the manufacturing but also in the service sectors. A remarkable technological forecast,[7] published in 1930, foresaw transatlantic jet air transport and expected ship owners to rush in and diversify in time. As we now know, the ship owners missed this chance of recognising the function 'intercontinental transport' (at different speeds and by different means), and are now left with the sole possibility of diversifying into the pleasure-cruising sector.

An A. D. Little study of the introduction of semiconductors[8] dramatically revealed how some of the most renowned and most powerful electrotechnical groups lost much momentum in the decisive early development and marketing phases because they placed the transistor, whose performance is, at first glance, comparable to that of an electronic tube, in their vertically structured tube divisions, where the new device was fought as 'the enemy' in an environment geared to product-oriented thinking.[9] The advantages of self-motivation which are created by giving a man a clear-cut operational problem to solve – 'Everybody loves his baby' – are now increasingly offset by the disadvantages due to a lack of strategic thinking. Product-oriented thinking is an obstacle to swift technological innovation.

Between step 2 and step 3 ('Shaping the Future'), there are a large variety of partial and transitory solutions which are coming more and more to the foreground in industry, in particular in the United States. In a company, this process of change often starts with the addition of a scientific or technical evaluation group (or Chief Scientist) to the corporate-level staff groups that deal with corporate long-range planning (in this context, frequently aiming mainly at five-year budget plans). This provides the possibility of looking at a potential development which would fall in between the vertical product lines, or to persuade one of the vertical 'empires' to adopt this area, or to set up new product lines, if necessary. In so far as such diversification is not just induced by the 'greener grass on the neighbour's lawn', it may be understood to some extent as a first

[6] A recent study of Soviet science policy, conducted by the United States Department of Commerce, revealed the astonishing fact that, in the 1930s, Stalin ordered the development of the Siberian hydropower sites because he thought that petroleum was too valuable to be burnt in power plants.

[7] Earl of Birkenhead, *The World in 2030 A.D.*, Harcourt, London 1930. The author was not a scientist or engineer.

[8] A. D. Little, Inc., *Patterns and Problems of Technical Innovation in American Industry*, Part IV: *The Semiconductor Industry*, Report PB 181573 to National Science Foundation, Washington, D.C., 1963.

[9] Today, function-oriented thinking would deal with semiconductors in the different functional areas such as communications (aiming at microminiaturisation, high frequencies, etc.); power transmission (aiming at high voltages for d.c. transmission, etc.); and power utilisation (specific heat capacity/electrical conductivity characteristics, etc.). The real goals and alternatives tend to become blurred in a unified, technology-oriented 'semiconductor' development programme.

move toward function-oriented thinking. Examples may be found in
forward and backward integration of product lines: e.g., building up
industrial chains from raw materials to end-products, such as the expansion
of the big oil companies in the direction of agricultural chemistry and
petrochemistry, and now further into textiles and foodstuffs, or the food-
processing industry integrating backwards to farming. A particularly
interesting backward integration is that from computer software to
computer hardware, which is actually taking place. Other types of 'functional' diversification are exemplified by the move of almost all important
aerospace companies into electronics.[10] Technological developments,
pushed to their extremes, often exhibit an inherent tendency to enforce

Product-Oriented Categories		Function-Oriented Categories	
Category	*Example*	*Category*	*Example*
—	—	Societal function	Communication, education, information, health, etc.
Industrial sector	Computer industry	Technological function	Information technology
Product line	Computer series	Functional capability	Data processing
Product (system)	Specific computer model	Technological capability	Speed, storage capacity, etc.
Components	Integrated circuits	Basic technology	Microminiaturi-sation, cryogenic memories, etc.

Table 8.2. A Typical Hierarchy of Categories.

integration, such as the 'growing together' of the electronic components
and systems sectors in the age of microminiaturisation, or of the airframe
and propulsion unit sectors for hypersonic planes (Mach 5 to 7, and
beyond).

The fact that a considerable fraction of European industry has not yet
implemented the shift from step 1 to step 2 – from horizontal to vertical
structure, in organisational terms – points to the possibility of 'leap-
frogging' from step 1 straight to step 3. On the outside, a function-oriented
structure does look like a compromise between horizontal and vertical
organisation, although there is a basic difference in attitude and in the
degree of penetration of the company. These organisational aspects will
be discussed in more detail in the following chapters.

Table 8.2 presents a rough idea of the principal categories in which

[10] The present trend of aerospace companies, both in America and Europe, to
diversify into the area of deep submergibles may be prompted primarily by the availability of special skill in these companies, but may develop into a bold functional concept
circumscribed by 'exploration – space and deep sea', or even 'conquest of environments
hostile to man – air, space, deep sea'.

thinking differs for step 2 and step 3. The more function-oriented thinking becomes, the more it will tend to incorporate all areas in which technology makes a contribution to the particular function. Only high-level functions, ultimately tying in with societal needs and goals, permit flexible control over the alternatives and true leadership.

In the United States, the Engineers' Joint Council stated this very clearly almost a decade ago in an attempt to look at the future:

> 'The fragmentation of most industries has led to concentration on materials and devices with little relationship to the technical and socio-economic systems within which these materials and devices must function. Only those industries that have large, integrated responsibilities, and that have been organised in the past few decades in co-operation with the public interest have developed effective systems that provide service to the general population.'[11]

The EJC report cites the communications field, power distribution, and fuel distribution among the positive examples.

The computer industry, for example, has failed, so far, to express its corporate objectives clearly in terms of high-level functions. 'Data processing' ties in with function-oriented thinking but only at a low level.

As the higher-order products assume the lead in product-oriented industry, the higher-order functions eventually will become increasingly dominant. Without any change in the attitude of the computer industry, the 'data processing' industry is inevitably bound to become the sub-contractor to 'information', 'communication', 'education', 'health' and other industries which are leading the innovation process. For maximum profit, this might not be considered a bad policy for quite some time to come; but maximum profit is not a challenging goal for professional people, as the 'minority leaders' of the top 10 per cent of U.S. industry (such as Bell Telephone Laboratories and the Xerox Corporation) testify. 'The corporate leader who does not try to conduct his company so as to instill pride in his people is doomed these days.'[12]

It should be emphasised that the managerial capabilities of steps 2 and 3 generally include those of the lower steps; e.g., a company focussing on the 'challenge' incentive will not neglect 'hedging against threats', and will be able to incorporate much of the momentum inherent in product-oriented thinking. The progression from step 1 to step 3 might best be taken as a widening of scope and horizon in overall management attitudes.

[11] Engineers' Joint Council, *The Nation's Engineering Research Needs, 1965–85*, Subcommittee Reports of the Engineering Research Committee, EJC, New York 1962.
[12] Joseph C. Wilson, *The Conscience of Business*, Westminster College, Fulton, Missouri 1965.

FUNCTION-ORIENTED 'INNOVATION EMPHASIS' IN INDUSTRY

As far back as 1935, S. Colum Gilfillan[13] observed that inventions of considerable social significance tend to arrive in clusters which are grouped around functions. This bears testimony to the power of social motivation which is inherent in technological innovation. The aim in industry should be to establish function-oriented thinking at the level of comprehensive control over the interactions of technology transfer. According to Harvey Brooks,[14] technology transfer is a complex interaction between vertical transfer (from a scientific principle to a technical realisation, etc.) and horizontal transfer (diffusion of existing knowledge, application and service engineering, inter-disciplinary fructification, consideration of 'commonalities', etc.).

The integration of technological forecasting and planning is inherent in this concept, not only because forecasting participates actively in the selection of goals and missions (as is already seen in product-oriented thinking), but also because the solutions offered by planning along alternative paths into the future interact, in a feedback cycle, with the evaluation of the alternatives which are anticipated through forecasting. While forecasting provides guidance for planning, its probability and desirability statements are, in turn, modified by the possible courses of action which are worked out by operational planning.

The difference between product- and function-oriented thinking may be demonstrated by a hypothetical example concerning the oil industry (see *Figure 8.2*). If it regards itself as 'oil industry' proper, i.e. concentrating on the resource oil and its derived products, a diversification policy characteristic of the development of big oil companies over the past twenty to forty years, up to the current research projects in the production of Single Cell Protein on a fossil hydrocarbon substrate, is the logical result (*Figure 8.2* (a)). If, however, a company feels responsible for the societal function 'transportation' (or, for simplicity of the example, 'automotive transportation'), it will (1) prepare for the switch to other primary energy sources; (2) evaluate the alternatives along with their significance to systems of human living (including the reduction of air pollution in urban areas and conservation of fossil hydrocarbons for upgrading rather than downgrading purposes), and probably familiarise itself with the idea of running systems of giant base-load nuclear reactors to convert primary nuclear energy to electricity for batteries or forms of fuel that are suitable for use in fuel-cells; and (3) possibly find an incentive even to participate actively in the development of future propulsion systems.

[13] S. Colum Gilfillan, *The Sociology of Invention*, Follett Publishing Company, Chicago, Illinois 1935.

[14] Harvey Brooks, 'National Science Policy and Technology Transfer', Conference on Technology Transfer and Innovation, 15–17 May 1966, Washington, D.C.

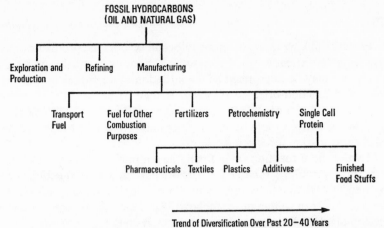

a) Product-oriented thinking exploits alternatives offered by the raw materials oil and natural gas. Diversification moves in direction of higher-order end products. (Actually, the diversification chain started long ago with lubrication and illumination products.)

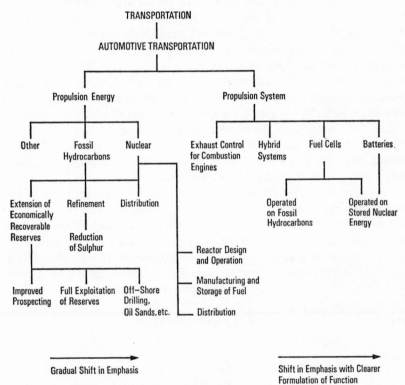

b) Function-oriented thinking moves toward a strategy based on primary nuclear energy. The original interest in propulsion energy is extended to include propulsion systems. New functions may crystallise around new skills (one big oil company, with particular skill in off-shore drilling, contemplates diversification into mineral collection on the sea bottom, and off-shore gold mining).

Fig 8.2   Impact of Product-Oriented and Function-Oriented Thinking on Innovation
(the latter is a hypothetical example).

Present thinking in big oil companies still favours the product-oriented approach, but already introduces some action, at a still lower level of effort, that may be interpreted as the initiation of a gradual shift to more prominent function-oriented thinking. It may be noted, in this context, that the big oil companies were among those industries which first recognised the incentive to join social research and forecasting to their technological and market forecasting.

Two or more functions may finally be adopted by the same company (e.g. 'transportation', 'food', etc.). The orientation toward functions – and the trend to fill functions more and more completely – may also favour mergers between different branches. The Esso/Nestlé partnership, to develop Single Cell Protein from petroleum, is an example of this trend. It represents a functional chain that is fully integrated, from oil exploration and drilling to the manufacture and marketing of finished foodstuffs.

Clearly, thinking in broad functions will be restricted somewhat where the alternatives demand too great a variety of skills. However, in the framework of high-level functions, companies have already expanded their interests into areas which nobody would have expected only a short time ago. Companies which are gradually aligning their thinking to an overall 'communications' function are acquiring publishing houses or youth periodicals (RCA and Xerox), or were even planning to enter city building (General Electric – 'the city as a communication network') – and ITT has acquired a big rent-a-car system. Companies which recognise an 'education' function for themselves perform research in new educational systems (e.g., Xerox, Raytheon, and computer companies). Airlines, taking their 'tourism-abroad' function more seriously, acquire or build hotel chains (Pan American, TWA, Air France, etc.).

Another restrictive factor, which may lead to a compromise in the establishment of function-oriented thinking, can be seen in a possibly large difference in the innovation rates that apply to individual branches of the same company. In the chemical industry, for example, the large 'base-load' share of traditional basic chemicals is changing very slowly, whereas the full drive for innovation is in progress in some of the end-product branches. A number of companies apply technological forecasting and conscious forward planning principally to the military part of their businesses, and that is where they contribute to rapid innovation.

A FUNCTION-ORIENTED FRAMEWORK FOR THE CHEMICAL
INDUSTRY

Particular questions which arise for the petroleum industry have already been mentioned above. Before implications for the chemical industry at large can be discussed, its present orientation, which reflects historical developments, must be clearly understood. The fact that an industrial sector can still be circumscribed by the name 'chemical industry' points to a particular and unique structure of industries basing their technology on a scientific discipline – a structure which is becoming an anachronism today. There has never been a 'physical' or a 'biological' industry, and all the so-called science-based industries outside the chemical sector orient themselves toward technologies (e.g. electronics or metallurgy), products (e.g. aircraft or computers), or functions (e.g. energy conversion or tele-communication). Many of the technological disciplines (such as mecha-nical engineering) never became the basis for an industrial sector whereas some disciplines that did – and the foremost example here is, perhaps, chemical engineering – emerged from the needs of existing industrial sectors and developed as a service to them.

The chemical industry started with the production of basic chemicals or the direct exploitation of scientific knowledge. Agricultural chemicals and pharmaceuticals (which had been a sector of some science-oriented crafts for many centuries) first extended the tasks of the chemical industry to finished products in two functional areas (food production and health) which could be unambiguously identified with certain lines of chemical product development at that time.

The picture became much more complicated with the development of finished products along two lines that inevitably led to collisions with other industrial sectors: plastics and synthetic fibres (or, in more general terms, man-made fibres including rayon). Whereas these developments were, at first, clearly meant to exploit scientific knowledge and special skills, the *invasion* of other industrial sectors – particularly the textiles sector – subsequently became a conscious objective and the source of many important and most successful innovations.

It is most interesting to note that the more traditional subsectors of the chemical industry did not show the same aggressive dynamics and to some extent have to defend themselves against powerful factors influencing their traditional roles. The agricultural chemicals sector, for example, has never arrived at a formulation of its tasks in terms of the 'food production' function and consequently never assumed leadership within this function. Not only is it today invaded by the petroleum companies (in the United States since about 1963) but it also has to take account of the influences exerted by the food industry due to the latter's 'backward integration' right through to the farmers and to some extent dictating the farmers' activity. The agricultural chemicals sector would, at this moment, be completely helpless had it to face any restructuring of the food production function, for example, in the form of the introduction of non-agricultural food production technology. On the other hand, the role of chemistry in food production is becoming ever more prominent whether in agricultural food production, in reducing food waste, or in the early attempts to supplement the old self-sufficient plant/animal/chemical fertiliser system (e.g. through chemically produced amino-acid additives and later through new biological-chemical processes to produce Single Cell Protein, etc.).

This example already shows that if the structure of the chemical industry could so far be called a mild anachronism, it will become a matter of life or death in the imminent future.

Two developments which change the nature of the scientific discipline 'chemistry' itself, tend to emphasise the incentive to abandon the discipline- or skill-oriented approach:

> Chemistry is becoming obscured, as a scientific discipline, by the increased understanding of phenomena at the sub-molecular level in terms of energetic principles, putting chemistry on the same basis as physics.
>
> Due to the understanding of biological phenomena at the molecular and sub-molecular level, the interface between chemistry and biology is quickly becoming the most important and most fertile area for new inter-disciplinary developments.

A look at *Figure 8.3*, which depicts a few of the functional technologies which are coming to the forefront now, confirms the rising importance of inter-disciplinary developments engaging in terms of original disciplines, chemistry and biology.

It is useful to attempt to group social functions and the corresponding functional technologies in line with the three basic types of human motivation as in *Figure 8.3*:

> (*a*) *'Hedging against Threat'*, which, through the centuries until relatively recently, has been the principal incentive for technological development since the early inventions to make hunting more effective and the beginning of agricultural technology some 5,000 years ago. The main theme – 'survival amidst scarcity' – not only continues to hold our attention due to the two types of imbalances created by the population explosion (giving rise to a dramatic world food problem) and the technological explosion (threatening the depletion of certain resources, such as fossil fuels), but is enriched by the potential destructive uses of technology and by the prospects of over-crowding, which constitute serious threats in themselves.
>
> (*b*) *'Maintenance of Equilibrium'*, meaning here, in general, the equilibrium of man's world within Nature, is only now gradually coming to our attention in its full importance. The interfaces between 'threat' and 'equilibrium' on one side – a serious disequilibrium may indeed become a threat to vital human interests (e.g. the population explosion, or pollution) – and 'equilibrium' and 'challenge' on the other side are not sharply defined. However, the focus of interest under this motivation is on areas where an equilibrium can be maintained or restored and where the basic alternative of substitution for our natural world (e.g. by creating wholly artificial environments or making other planets or the oceans habitable) do not yet have to be taken into account as inevitable.
>
> Factors disturbing the equilibria are again primarily due to the two explosions in population and in technology utilisation. The distortions of the

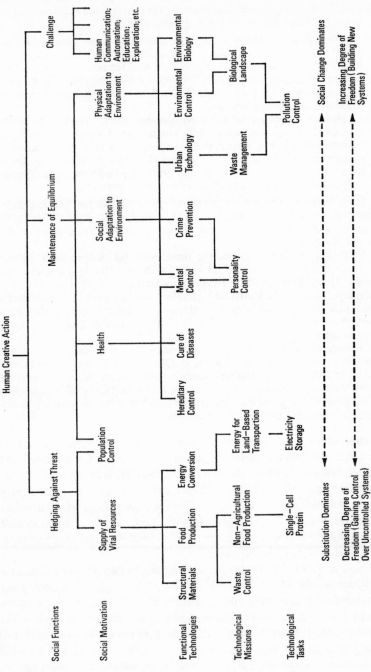

Fig 8.3 Examples from the technological decision-agenda in relation to social functions and basic human motivations. The representation by a relevance tree depicts only the primary relationships. Numerous inter-relationships and 'commonalities' exist on all levels from functions downwards (e.g. the biological landscape depends in an important way on the future share between agricultural and non-agricultural food production). The three motivations also overlap to a large extent.

human environment have become obvious and many of them are irreversible. New stable or slowly changing patterns of the man–Nature relationship may be conceived which involve changes on both sides:[15]

> Changes in environment through creating a new 'biological land-scape' which would be situated somewhere in between conserved (wild) nature and the spreading urban and industrial landscapes of today. This task, taken in a broad sense, is inter-related with the planning of urban settlement, with transportation, food production, and other functional technologies affecting land use.

> Human adaptation to new environments which may be seen as the principal function of a future health programme. Another disequilibrium concerns the effects of eliminating, to a large extent, the natural selection process for the human race suggesting the necessity of new modes of hereditary control. Molecular biology provides us with the understanding of biological processes which are becoming controllable.

> (c) *'Challenge'* is to a large extent a type of motivation that is characteristic of advanced countries which have essentially overcome the problems of scarcity. 'Vigorous societies harbour a certain extravagance of objectives' (Alfred North Whitehead).

> Challenge has also become the most powerful driving force behind economic development, above all in the areas marked by advanced and rapidly advancing technology. The human nature, originally, was not that of a worker, but that of a hunter and gambler (Nigel Calder). It will become a major task for the social sciences to elucidate the deeper layers of this motivational complex 'challenge' which is also the creative source for every cultural development.

> The spectrum of topics recognised today – from automation to exploration, from universal communication (a type of exploration of the human world) to continuous education – seems to bear out the pattern of timeless instincts striving towards freeing man from the necessity to work, domination of environment (the old hunting instinct), and shaping his own anthropomorphic world (the core of human culture).

It can readily be seen that the 'natural choice' of functional technologies which the present chemical industry could adopt for its functional objectives, fall mainly under the two motivations 'Hedging against Threat' and 'Maintenance of Equilibrium'. This means that the chemical industry will be in the foremost frontier in pushing back the threats which are now being recognised and that its operations will become part of a world-wide 'crisis management', especially in the areas of food production, population control, and environmental control.

Of course there will be prominent chemical tasks in the traditional sense which would fall under the motivation 'Challenge'. One needs only to remember the important role chemistry already plays in the electro-technical industry, in communication, in space exploration (chemical rocket-fuels, etc.). Some of these tasks will perhaps always stay with the present chemical industry. However, the 'functional lead' will be with other branches of industry. If, for example, important breakthroughs in

[15] See also Chapter 3 of this volume.

chemistry are expected to contribute to information technology (which comes primarily under such functions as automation, communication, and education) – information storage through colour changes, for example – this does not mean that such innovations will, or should, be developed by the present chemical industry.

Already the aerospace industry is developing materials (chemical or metallurgical) through its in-house R&D effort and having them produced outside according to precise specifications. Originally 'chemical' goods, such items as computer tapes and video tapes are no longer left with the chemical industry.

Whatever the extent to which the present chemical industry is able to act as a supplier to 'prime-contractors' in other functional areas, the crucial question will be to find those functional technologies in which it can lead and plan in an integrated way. This will necessarily lead to a considerable invasion of other traditional sectors. A recent joint-sponsorship study of the Battelle Memorial Institute, 'Plastics in House-building', found sponsors only among the chemical industry, not among the building industry. The same area of urban technology, however, will not be an uncontested prey for the chemical industry – the interest of big electro-technical groups (regarding the city as a communications network) is well known.

If the chemical industry re-defines its primary objectives in terms of functions, falling under the motivations 'Hedging against Threat' and 'Maintenance of Equilibrium', of all the industrial sectors it will have least to worry about the future needs to be served. It will also be in a position where its services to humanity will be more readily accepted since the aspect of social change is overshadowed by the need for substitution.

But in attempting to gain back control over uncontrolled systems in a stage where serious disequilibria are already making themselves felt, the present chemical industry will have to do the most sophisticated planning for strategic decision making and policy making that will be demanded from industry. It will be on the control levers of a global 'crisis management' which will characterise the coming decades.

If chemical engineering, in a broad sense, is circumscribed as the engineering discipline dealing with any type of production process for the present chemical industry, it will follow the chemical industry and serve it also in the emerging function-oriented framework. One may go further and assume that the emphasis on new developments in chemical engineering will be determined to a decisive extent by the functions the dynamic chemical industry adopts for its future objectives.

The most important extension of the scope of chemical engineering will concern biological processes. Whereas biochemical and biomedical engineering are at present still based primarily on the impact of chemical processes on living matter, biological processes will have to be mastered in future in the same way as chemical processes are now. If chemical engineering is to optimise and control biological processes in the same way as it now handles chemical processes a whole body of macroscopic theories of statistical behaviour ought to be developed. This includes, for example, detailed knowledge of reproduction, response to environmental conditions, etc., of a wide variety of micro-organisms. The step from microscopic scientific knowledge to macroscopic principles guiding chemical engineering may be compared to the step from the knowledge of physical phenomena in boundary layers to the dimensionless, handy formulae of heat and mass transfer which make it possible to optimise complex multistage processes. In addition, the unique possibilities of utilising self-optimisation and self-control – which one might call the cybernetics of biological processes – and going far beyond the equivalent possibilities in chemical processes, will have to be brought into the play.

The spectrum of new tasks for chemical engineering may be briefly sketched by using the example of food production technology, in which economic optimisation is a particularly stringent factor. One may imagine new tasks, for example, in the following areas, some of which are already tackled on a laboratory or industrial scale:

> Cheap chemical and biological production of amino-acids for additives;
> Multistage production processes for Single Cell Protein on various substrates (crude petroleum or paraffins, organic wastes, forest wastes – the latter two also involving economic collection and processing techniques);
> Process chains from Single Cell Protein or other forms of protein concentrate (e.g. from plant proteins) to finished food products;
> Single Cell Protein production from algae in water tanks;
> Large-scale photosynthesis in non-living matter (achieved on a laboratory scale so far);
> Non-photosynthetic food production (using chemical or other energy input), up to the possibility of non-agricultural tissue cultures;
> Storage and re-deployment of light; etc.

There is a vast variety of new tasks also in other functional areas, for example storage of electricity for the purposes of transportation (e.g. in liquids suitable for use in fuel cells, etc. – although the use of fuel cells seems more remote at present than other forms of electric transportation), the economic production of prefabricated plastic houses, the large-scale reprocessing and desalination of water, and the spectrum of opportunities presented by biomedical engineering.

One may even wonder whether the present scope of chemical engineering may be extended not only by the addition of 'biological engineering' but also by software disciplines which deal with the application of technology to social needs with the diffusion and adaptation of technology and with the optimisation of production processes and product introduction in different social environments. Enrichment by the social sciences would put chemical engineering in a better position to optimise its contribution to the flexible functions which constitute the response to 'Hedging against Threat' and 'Maintenance of Equilibrium'.

In any case, chemical engineering is emerging as the key element in the attempt to find *economic solutions* to the problem of substituting for Nature in areas which are becoming most crucial on a global scale, i.e. areas where threats to humanity have to be met. The extent to which these solutions will be applicable to the problems of the developing countries will determine the success or failure of a global 'crisis management'.

A TECHNOLOGICAL DECISION-AGENDA FOR THE FOOD PRODUCTION FUNCTION

Food production, although becoming of increasing concern in a global context, is still locked into specific technology- or product-oriented thinking. However, it is becoming increasingly doubtful that boosting agricultural productivity, or the efficiency of ocean-harvesting, can, in the long run, cope with population increase or constitutes a wise strategy at all.[16] The following example indicates how a technological decision-agenda for the overall human food production function may be systematically developed.

At present, there is a calorie 'gap' in the world, but apparently a relatively marginal one (perhaps not more than 6 to 8 per cent globally, with 20 per cent as the figure for countries like India).

However, the predominant aspect of the world food problem today is the shortage of animal proteins – or, more accurately, of amino-acids in a blending which is representative for animal proteins. The full importance of the nutritional aspect has been brought to the foreground by research during the past ten years only. Research has clarified the roles of the different amino-acids and their blending in human food: Out of the twenty amino-acids man needs, the human body is capable of synthesising twelve, whereas the remaining eight have to be taken in directly with food. The human body uses them only in a specific blending – if, in food protein, only one of these amino-acids is present in less than the proportion required, all the other amino-acids are utilised only partly, at a fraction that corresponds to the level at which the required proportions for the blending are fulfilled. (This is a simplified picture; we know well on what proteins rats grow well, but not yet fully what man needs.) Recognition of this gave rise to the concept of different 'biological value' of proteins. The importance of animal proteins in human food becomes clear on this basis. At the same time, this understanding provides us with the necessary knowledge to compose a proper diet by combining amino-acids

[16] See also Chapter 14 of this volume.

from different natural or synthetic sources, and to improve the biological value of natural proteins (e.g. plant proteins) by adding such amino-acids in which they are deficient.

On the basis of classical agricultural food production, the average daily demand for pure protein in human food is approximately 70 to 75 grams, out of which 30 grams ought to be animal protein (meat, fish, eggs, milk). In the United States, the average man has approximately 95 grams per day at his disposition, whereas the figure for India – 45 grams per day – indicates a severe lack of protein in general, not to speak of the lack in animal protein.

Today, approximately 20 million tons of pure animal protein are available per year – 17 million tons from agriculture, and 3 million tons from ocean harvesting – whereas 34 million tons would be needed. This is the real world food problem today.

Agriculture produces approximately 150 to 180 million tons per year of pure plant protein, out of which 80 million tons are eaten by man (40 million tons in form of cereals), and the rest by animals. The 'conversion efficiency' from plant to animal protein is approximately 15 per cent (other sources claim that it is only 8 to 10 per cent, which would not be fully consistent with the figures quoted above). If the present 'gap' of 14 million tons of animal protein per year was to be filled by more animal breeding within the classical agricultural system, the corresponding increase in plant protein production would have to be at least 100 million tons per year. This is considered as not feasible under the given circumstances. For the future, the requirements would become even more stringent.

It can be seen from the above figures, that there would be, at present, no 'gap' in the total protein consumption on a global basis – the total protein demand is about 100 million tons per year – if this protein was in the proper blending. However, it is not, and the most severe problem arises from the lack of animal protein which represents only $\frac{1}{9}$ of the calorie intake at present. Ayres[17] claims that full utilisation of the resources, including waste forest products (accounting for 40 per cent) could raise this fraction to $\frac{1}{3}$. If new protein sources become available that could partly replace animal protein from agriculture, this would not only contribute to filling that particular 'gap', but also relieve agriculture partly from the burden of 'inefficient' animal breeding – and thereby ease the calorie problem.

In man's history the following sequence of steps in the development of a food production system may be distinguished:

1. Wild growth (mainly terrestrial, supplemented by ocean), i.e. food gained by hunting and collecting, had to be abandoned when population density became higher than 1 man per 10 square kilometres.

[17] Robert U. Ayres, *Technology and the Prospects for World Food Production*, Hudson Institute, Report HI–640–DP (Rev. 2), Harmon-on-Hudson, 19 April 1966.

2. Mixed plant/animal agriculture (supplemented by some ocean harvesting) on a self-sufficient basis, i.e. soil fertilisation by animal dung.

3. Extension of mixed plant/animal agriculture by soil feeding with non-biological fertilisers (supplemented by ocean harvesting pushed to half the natural limit).

4. Supplementation of the preceding system by concentrated protein additions, introduced now (with proteins from agricultural plant and ocean food sources).

Within the existing food production system, the following protein sources are yet practically untapped for human consumption: Oil seed cake (containing 18 to 20 million tons of pure protein per year, at present), fish-meal (perhaps of the order of 1 million tons per year), and yeasts grown on molasse and agricultural waste (0·2 million tons per year). The agricultural system could easily be adapted to growing leaves to increase the protein production. Apart from oil seed cake, only yeasts could conceivably contribute on a large scale to protein production.

Ocean harvesting (of wild growth) can be increased by a factor of perhaps 2·5, not more. The existing agricultural system can be expanded by new plant and animal species, higher yield, and new land – how much, is a most controversial issue – but only by tedious and gradual change, which is believed to be insufficient for meeting the problems as we approach the turn of the century.

Therefore, it is likely that the present food production system will have to be expanded so as to include new approaches. A systematic survey of options may be undertaken on the basis of a *morphological analysis of the food production function* (*see Table 8.3*).

Step 1 of the above sequence, i.e. primitive man's system, depending entirely on wild growth, included only four possibilities:

A1 – B1 – C1 – D1 – E1 – F1 – G1 – H1: Collecting wild fruit, etc.

A1 – B1 – C1 – D1 – E1 – F2 – G1 – H2: Hunting wild terrestrial animals and birds

A1 – B1 – C2 – D1 – E1 – F2– G1 – H2: Fresh water fishing

A1 – B1 – C3 – D1 – E1 – F2 – G1 – H2: Ocean harvesting (fishing, collecting other seafood, etc.)

All four options continue to be currently exploited, but are negligible in their contribution to the present food production system with the exception of ocean harvesting which has already been pushed to half of the natural limit.

	1	2	3	4	5	6
A	*Energy source*	Solar	Fossil (stored solar)	Nuclear			
B	*Primary energy conversion process*	Photo-synthetic	Non-photo-synthetic				
C	*Environment*	Terrestrial	Fresh and brackish water	Ocean	Industrial		
D	*Environmental conditions*	Natural	Controlled natural	'Augmented' natural	Artificial		
E	*Primary production process*	Wild growth	Culture	Industrial			
F	*Form of primary nutritional matter*	Plant	Animal	Micro-organisms	Organic tissue	Natural organic resources	Natural inorganic resources
G	*Food processing for human consumption*	Direct consumption	Extraction	Controlled biological transformation	Controlled chemical transformation		
H	*Basic form of food for human consumption*	Plant	Animal (meat, milk, eggs, etc.)	Micro-organisms	Organic tissue	Separated basic food substances	

Table 8.3. Morphological Analysis of the Food Production Function, developed here for a maximum of two major transformation steps (e.g. solar energy into plants, plants into animals). The 34,560 different possibilities which the analysis yields, may be reduced to 3,168 'realistic' possibilities after the elimination of internal inconsistencies and major impracticabilities. Only 18 possibilities are currently in use, and only 5 of them play a major role in the current food production system for human consumption.

Step 2 of the sequence introduced additionally agriculture on a self-sufficient basis, as well as some fresh and ocean water culture:

A1 – B1 – C1 – D2 – E2 – F1 – G1 – H1: Agricultural plant production

A1 – B1 – C1 – D2 – E2 – F1 – G3 – H2: Agricultural animal raising
(for meat, milk, eggs, etc.)

A1 – B1 – C2 – D2 – E2 – F2 – G1/3 – H2: Fish ponds

A1 – B1 – C3 – D1/2 – E2 – F2 – G1 – H2: Mariculture (oysters, etc.)

The two agricultural possibilities continue to play an important role, but are naturally limited. Mariculture seems to hold a tremendous potential for the future, the realisation of which will require major technological developments (generally considered to be a few decades away).

Step 3 introduced artificial conditions for agriculture, mainly through non-biological (chemical) soil fertilisation:

A1 – B1 – C1 – D3 – E2 – F1 – G1 – H1: Agricultural plant production
with chemical fertilisers

A1 – B1 – C1 – D3 – E2 – F1 – G3 – H2: Agricultural animal raising

A1 – B1 – C1 – D4 – E2 – F1 – G1 – H1: Agricultural plant production
under artificial conditions
(temperature and light control
in greenhouses, etc.)

Step 4 has led so far to the marginal exploitation of the following possibilities to concentrate protein:

A1 – B1 – C1 – D2/3 – E2 – F1 – G2 – H5: Protein extraction from
vegetables (e.g. Incaparina),
or oil seeds (so far, human
consumption negligible)

A1 – B1 – C1 – D2/3 – E2 – F1 – G3 – H3: Yeasts, grown on molasses
(so far little for human
consumption)

A1 – B1 – C3 – D1 – E1 – F2 – G2 – H5: Fish-meal (recently introduced for human consump-
tion)

A2 – B1 – C4 – D4 – E3 – F5/6 – G4/5 – H5: Industrial production of
synthetic amino-acids

T.P.A.S.F.—K

The development of the following options is in an experimental stage, as far as human consumption is envisaged:

A1 – B1 – C1 – D1/2 – E1 – F1 – G4 – H5: Carbohydrates from wood (experiments after World War II discontinued due to lack of incentives)

A1 – B1 – C1 – D1/2 – E1 – F1 – G3 – H3: Yeasts, grown on organic waste

A1 – B1 – C1 – D3/4 – E2 – F1 – G2 – H5: Protein extraction from leaves

A1 – B1 – C2 – D3/4 – E2 – F3 – G1 – H3: Fresh water algae cultures

A2 – (B1) – C4 – D4 – E3 – F5 – G3 – H3: Yeasts and bacteria, grown on fossil macromolecular substrates (petroleum)

Some of the food production processes including the factor H3, i.e. processes leading to scp (Single-Cell Protein (yeasts, bacteria, and single-cell algae), are already used at medium to large scale for animal feeding. They represent very important future options also for human consumption.

Some of the major options for the future may be seen in the following configurations:

A1 – B1 – C1 – D1/2 – E1 – F1 – G2 – H1: Pines and conifers, partly edible after detoxication

A1 – B1 – C1 – D3/4 – E2 – F1 – G2 – H5: Growing of special leaves for protein extraction

A1 – B2 – C1 – D3/4 – E2 – . . . : Non-photosynthetic production of leaves, yeasts, algae, and possibly tissue cultures (with solar heat as the major energy input)

A1 – B1 – C3 – D2/3 – E2 – F2 – G1/3 – H2: Large-scale mariculture (fish and seafood)

A1 – B1 – C3 – D2/3 – E2 – F3 – G1 – H3: Ocean algae cultures

A2 – (B1) – C4 – D4 – E3 – F5 – G4 – H5: Chemical protein synthesis from fossil (macromolecular) resources

A2/3 – B1 – C1/4 – D4 – E2/3 – F1/3/4 – G1/2/3 – H1/2/4/5: Terrestrial food production on basis of artificial photosynthesis (feasibility doubtful)

A2/3 – B2 – C1/2/4 – D4 – E2/3 – F1/2/4 – G1/2/3 – H1/2/4/5:
> Terrestrial or industrial food production on basis of non-photosynthetic energy conversion (not practical at large scale, or only with nuclear fusion energy?)

Apparently, the biggest potentials can be ascribed to the following technologies: Biological transformation of fossil macromolecular resources (petroleum, natural gas) yielding Single Cell Protein (SCP) with a production potential many times the total protein demand; photosynthetic and – to a higher extent – non-photosynthetic production of protein-rich algae, yeasts, tissues, etc., under artificial environment conditions (for example, in water tanks); mariculture; and transforming the forests of the world chemically (yielding only carbohydrates) or biologically (protein-rich yeasts).

An ultimate, but very remote alternative may be seen in the possibility to change the human genetic code in such a way that man can get along without animal protein at all (by creating a new human capability of synthesising also the eight amino-acids which have to be provided directly now). However, such a use of science is certainly beyond our ethical grasp today.

Research is concentrating mainly on increasing agricultural productivity, and so far tends to neglect (with the exception of small-scale laboratory work) the essential alternatives. However, a decisive breakthrough is becoming visible already today in one of these alternative solutions: the production of Single Cell Protein from petroleum.

Another important consideration enters here: It has been assumed that agricultural productivity (per caput production) will increase more rapidly in the developed countries than in the less developed countries. The conclusion is drawn[18] that food exports from the developed to the less developed countries will increase, and that the less developed countries will have to improve their economy mainly by industrialisation to be able to import food. However, purely industrial (non-agricultural) food production technologies can lead to a compromise in building such food producing industries in the less developed countries. It is believed that such technologies could be introduced on a large scale in the second half of the 1970s. Single-Cell Protein production may soon become a simple industrial process which can be set up, for example, in any country where petroleum refineries are economic (i.e. where petroleum is found, or can be transported cheaply by means of pipelines or supertankers).

[18] Thorkil Kristensen, *The South as an Industrial Power*, internal OECD Document, 1967.

The implications for land-use, generally, become clear by comparing the following 'efficiencies': Whereas one cow, weighing 500 kilograms, produces 0·5 kilogram protein per day (both in form of milk and meat), the same mass of micro-organisms are capable of producing 1,200 kilograms of protein per day (by virtue of their rapid reproduction cycle).

ORGANISING FOR THE FUTURE

Institutional and organisational structures are closely inter-dependent with the types of planning carried out in them. Whereas currently the tasks of forecasting and long-range planning may still be looking for a place in existing structures, it will increasingly be the long-range perspective on society's future which will determine organisational forms.

For the cybernetic processes comprised in the notion of 'futures-creative' planning, *flexibility* becomes the key notion for organisation design. Thus, the evolutionary trend is leading from instrumental and pragmatic to adaptive institutions and inter-institutional patterns. An adaptive institution is capable of reflecting on its own *role* in a changing environment and of reformulating its policy accordingly. It also stimulates the generation of strategic alternatives and provides a framework for flexibly dealing with them.

The creative inputs needed for this internal and external organisational flexibility and adaptibility can only be expected if the principle of *decentralised initiative and centralised synthesis* can effectively be put to action. A new theory of multiechelon co-ordination may provide the basis for new approaches to organisation theory, relevant for this task.

Chapter 9 formed a presentation to the OECD Working Symposium on Long-Range Forecasting and Planning, Bellagio (Italy), 26 October–2 November 1968. It was first published under the same title in: E. Jantsch (ed.), *Perspectives of Planning*, OECD, Paris, 1969, and is reprinted here by permission of the Organisation for Economic Co-operation and Development (OECD), Paris, which holds exclusive copyright.

Chapter 10 is based on a presentation at the conference on Technological Forecasting: Concepts and Methods, organised by the Industrial Management Center (Prof. James R. Bright) on Hilton Head Island, South Carolina, 8–12 June 1969. It was first published under the title 'New Organizational Forms for Forecasting' in *Technological Forecasting*, Vol. 1, No. 2 (Fall 1969), and is reprinted here by permission of American Elsevier Publishing Company, Inc., New York.

Chapter 11 was first published in form of a book review in *Technological Forecasting and Social Change*, Vol. 3, No. 1 (1971), and is reprinted here by permission of American Elsevier Publishing Company, Inc., New York.

9. *Adaptive Institutions for Shaping the Future*

THE LINK BETWEEN PLANNING AND THE INSTITUTIONAL FRAMEWORK

PLANNING and institutions are mutually inter-dependent. A specific type of planning demands, and makes possible, specific types of institutions, and vice versa. The example of central planning at national level, as it is practised in Eastern countries, demonstrates the spectrum of institutions which are directly affected, although varying widely in their scope. Extending far beyond the institutions set up to elaborate plans, this spectrum also comprises the institutions set up to implement plans – as a matter of fact, most of the institutions and instrumentalities carrying out tasks in relation to the needs of society. One may go further in stating that a specific type and 'spirit' of the planning process correspond to a specific structure of or within society which, in turn, finds its expression in particular types of institutions.

If we take planning in the sense of *integrative planning*, cutting across a multitude of dimensions implicit in the interaction of Nature, man, society, and technology[1]– as well as factors and sub-categories such as economics and politics – the design of plans and of institutions has to be seen as an integral process, not as a parallel or consecutive process. Integrative planning is planning for change in a complex dynamic system, i.e. planning for a specific outcome, not just of action or elements (e.g. new technologies) leading to change. It includes measures to implement this change as well as to live with it subsequently in a systems framework. The *planning of institutions* of both types – for the realisation of planned change, and for subsequent control – is an important aspect of integrative planning. This means, for example, that planning for technological innovation ought to include planning for new social institutions.

In a world characterised by evolution toward greater integration, the scope of planning has inevitably extended and with it the types of institu-

---

[1] See Chapter 3 of this volume.

tions required to design and operate plans. As a consequence, institutions with international and world scope are becoming mandatory, today, for many problem areas transgressing national or regional boundaries. Suffice it to name here the World Food Problem.

INSTITUTIONAL REQUIREMENTS SET BY 'FUTURES-CREATIVE' PLANNING

Our perception of the evolving necessity to establish 'futures-creative' planning (Ozbekhan) is still primarily intellectual and still basically lacking the realistic angle of view which could give us a clear concept of desirable institutional changes. Alas, this seems to be inherent in human thought. Throughout history, institutional change has seldom kept pace with the acquisition of new insight, but has taken place as a delayed response to increasing pressures (or even major crises and catastrophes) resulting from environmental change. It appears justified, therefore, to derive the institutional requirements from a planning model first, although the development of planning and the proper institutional framework is in reality an integral feedback process.

Our present situation can be diagnosed as the beginning of a major crisis, which is developing in a period measuring in years and decades rather than over the centuries, as in the history of mankind to date. However, for the first time, we can use long-range forecasting and planning as a tool to study the consequences of insufficient or delayed institutional change. We ought to use this newly acquired capability to apply spur, guidance and direction to change – even now it is too late to avoid the crisis, but it could still contribute to averting the catastrophe.

Looking at the evolving concept of 'futures-creative' planning, we may readily recognise a few key characteristics which ought to determine the types of future institutions. 'Futures-creative' planning is, *inter alia*:

—*integrative*, implying the need for institutions to plan and operate on a 'system-wide' basis (Ozbekhan). The 'systems area' varies according to objective and handling capability, but planning for change in a complex dynamic system – as we described integrative planning above – must involve simulation on the basis of multivariate inputs. Characteristically, social, political, economic, and technological aspects of problems pertaining to the joint systems of society and technology must be dealt with.

—*normative*, implying the introduction of 'a new kind of leadership indicating distant goals of such large significance that they dominate the trivial trends towards expansion for expansion's sake'.[2] The leadership aspect is of crucial importance. Before we can apply comparative evaluation to prepare a rational choice, we need creative inputs at many levels. We need institu-

[2] René J. Dubos, 'Scientists Alone Can't Do the Job', *Saturday Review*, 2 December 1967.

tions to conceive of and study complete anticipations (intellectively constructed models of possible futures) and their functional elements, to devise alternative strategies – provide a rich 'decision-agenda' (K. Boulding) – and to point out clearly consequences for the future as well as for the present. We need institutions to assess and forecast value dynamics, and institutions to establish and sharpen consensus where appropriate.

—*adaptive* (one may even say 'cybernetic'), implying – for the medium- and long-range end of planning – the necessity for institutions relating continuously planning objectives and potentials to the changing environment, and to provide suitable information systems. However, capacity to adapt also requires a high degree of flexibility in the implementation of plans – implying, for complex problems, quite new and flexible types of institutional and inter-institutional patterns.

From an organisational angle, two important requirements should be pointed out which are partly implicit in the 'spirit' and the mechanisms of the planning model summarised above. The 'futures-creative' planning process should be:

—*democratic*, i.e. based on decentralised initiative and centralised synthesis, requiring effective communication to permit full understanding of corporate policies and objectives, and thus encourage self-motivation at all levels and stimulation and guidance of creativity.

—*not responsible for decision making*, but providing rather the full information base for decision making in a systematic manner. Planning aims at rationalisation of the *basis for action*, and not at the rationalisation of the action itself. This implies an organisational requirement for the interaction of planning and decision making at the proper levels and at the proper decision points.

THE EVOLUTION TOWARD ADAPTIVE ORGANISATIONAL CONCEPTS

It is useful to look back for a moment at the changes in organisational concepts over the past few decades. This will help us to understand the changes which are now required, as evolutionary rather than revolutionary.

We may distinguish between three basic types: *instrumental, pragmatic*, and *adaptive* organisation. These types hardly ever appear in pure form, but as combinations in which one type dominates. A fourth approach is suggested today by the revival of the anarchist ideal of purely individual self-fulfilment in conjunction with the abolishment of institutions.

Instrumental organisations are primarily geared to the deployment of more or less rigid sets of material and non-material resources for innovation (or conservation) and not to the innovation process as such. They may be regarded as semi-mechanical instruments capable of playing one tune only, but in different keys in the absence of well-defined targets, and ignoring completely strategic alternatives. The tune is provided mainly by tradition. Instrumental organisations preserve linearity of action, are insufficiently sighted on future objectives and outcomes and attempt systemic consistency mainly with regard to quantitative problems of resource deployment, such as working out distribution rules for resource allocation against a background of traditionally established patterns of values, merits, and various diffuse 'cultural' factors. In other words, instrumental organisations tend to apply a pseudo-rational 'decisionistic model' which may be considerably influenced by the interaction of pressure groups. Frequently, instrumental organisations are characterised by the absence of planning altogether.

The structure of instrumental organisations is usually horizontal and is defined in categories such as scientific and technical disciplines, skills, industries, armed services (Army, Navy, Air Force) and administrative entities. These categories refer back to the instruments themselves and tend to be captive to them; they are incongruent with the categories applying to objectives and outcomes, and the systems they form; therefore, they cannot deal with the quality of the outcomes. The structure of instrumental organisations is further characterised by a broad 'budget-centre' concept.

Examples of instrumental organisations in the public domain are Science and other Ministries, and National Science Councils. A clear case is the U.S. National Science Foundation. Although much of the traditional scope of ministries or governmental departments and agencies belongs to the instrumental type, the insufficiency of this resource-oriented rather than action- or target-oriented approach becomes particularly obvious in the institutions dealing with scientific and technological innovation.

In the military area, the initial and absurd 'rocket race' between the different American armed services may be remembered, to which the introduction of the Planning-Programming-Budgeting System put an end.

In industry, the horizontally integrated hierarchic pyramid, still prominent in Europe, typifies the instrumental approach. It was appropriate for the 19th-century-type entrepreneur who planned, invented, and decided everything himself, and for his classical capitalist attitude, but cannot cope with the problem of rapid technological change faced by the big anonymous companies of today. Diversification policies based on the exploitation or deployment of material and nonmaterial resources (raw materials, special skills, etc.) also are typical for an instrumental industrial environment.

In higher education, finally, the inherited structure of the university and its traditional faculties may be called instrumental. However, the general cultural aspect of the original *universitas* (rarely manifest today in university curriculae) represents values at a higher level. The broad instrumental character of some of the fundamental research organisations, especially those in 'big science', has encouraged the linking of fundamental research to higher education, and has thereby reinforced the application of instrumental criteria to both areas.

Central planning, which, essentially, is usually little more than production scheduling and a quantitative economic calculus, may be considered as pushing the instrumental approach to the extreme. A number of qualitative refinements, introduced, for example, into the French National Plan, do however provide a certain flexibility of this concept.

Pragmatic organisations are geared to action leading to well-defined goals, usually accepting a medium-range look into the future (as far as the 'freezing' of such goals permit). They are not, or are little, concerned with defining the goals of such action. In general, a recognised goal corresponds to a specific strategy. Pragmatic organisations may therefore be regarded as *ad hoc* arrangements for effective tactical (operational) planning and implementation. They become a problem when they tend to become permanent (as it is normally the case, encouraged by and contributing to society's inertia).

In technological innovation, pragmatic institutions are set up to develop specific and pre-defined technologies, acquiring and grouping resources in an optimised way. Pragmatic organisations have had spectacular success when applied to tasks of an inter-disciplinary character, particularly in areas of advanced technology. Frequently, such inter-disciplinary set-ups are referred to as 'mission-oriented' organisations, which is not quite correct if there is no built-in flexibility to play with more than one strategic option.[3]

Pragmatic organisations have been vehicles for the enormous acceleration of technological, and thereby social, change in our time. But it cannot be denied that they are also, to a large extent, responsible for the dangerous divergent trends in the viable system on our planet. Pragmatic organisations tend to push certain strategies to their extremes, while neglecting alternative options. Like instrumental set-ups, they function through linearity in planning and action – although deploying alternative sets of resources more flexibly than those; they aim at quantitative, not qualitative, targets. 'Expansion for expansion's sake' finds a propitious framework in pragmatic organisations, as does the single-minded profit maximisation motive, and the materialistic goals of the crude consumer society.

Lacking a clear view of the consequences of their work, pragmatic organisations are also primarily responsible for messing up the environmental system in which man has to live, in particular the joint systems of technology with man, society, and Nature. Moreover, they produce a certain

[3] The broad mission-oriented agencies of the U.S. Government, for example – primarily AEC (Atomic Energy Commission) and NASA (National Aeronautical and Space Administration) – exhibit both instrumental and pragmatic features (e.g. the Apollo programme, handled in a highly pragmatic way), but in this combination already inclined toward the adaptive type, especially NASA.

distortion of social functions and objectives. The 'polarisation' of higher education to serve the needs of industry – today representing the most pragmatic organisation of them all – is already becoming of major concern. There would be nothing wrong with such a 'polarisation', if industry's role were to be seen in a wider social context – but this already leads towards a new type of adaptively organised industry, which is only beginning to emerge now. It is also in the framework of pragmatic industry that Galbraith's[4] warning of the dictate of the 'technostructure' acquires real meaning.

The characteristic structure of a pragmatic organisation is one of strong vertical 'columns', or even relatively independent 'empires' with well-defined boundaries, thereby encouraging human ambition which is challenged by the prospect of power and individual success. The 'budget-centre' approach still prevails here, but on the narrow basis defined by such an 'empire'; from a corporate point of view, giving the vertically defined units financial responsibility, constitutes already a first step toward the 'profit-centre' concept.

Much of industry, especially in the United States, is today organised in the pragmatic form. Its primary characteristic is a structure composed of vertical product and service lines which are pushed to the extreme and 'defended' against alternatives which might better satisfy simultaneous systems criteria. Non-organic diversification policies, leading to the so-called 'conglomerates' grouping heterogeneous product and service lines mainly in order to build up financial power, may also be considered to belong here.

Pragmatic governmental agencies have grown mainly in the military and space areas and have developed sophisticated forms of project management.[5] However, it should be noted, that certain civilian parts of government also have pragmatic character, e.g. Ministries for Agriculture or Fisheries (rather than Ministries for Food Production and Supply), or Railroad Boards (instead of Ministries of Transportation, a trend which is now gaining ground).

In higher education, the prototype for the pragmatic approach is the modern Institute of Technology, with its technology-oriented departments, which have had striking success in newly developing inter-disciplinary areas while, at the same time, permitting a deeper penetration into sharply defined scientific and technological disciplines (which may also be more or less pragmatically defined, such as aerodynamics, or microwave communications). The general aim of the pragmatic university or Institute of Technology is the generation of specialists. Specialists, in turn, generate

---

[4] John K. Galbraith, *The New Industrial State*, Houghton-Mifflin, Boston 1967.

[5] It is significant that, in the United States, the Pentagon-controlled laboratories are only now gradually changing from instrumental to pragmatic types, whereas military contract policies for industry have played a major role in sharpening the pragmatic concept of industry.

sequential solutions which clog the systems with which man and society have to live.

As has already been said, the advance from instrumental to pragmatic organisations has been marked by spectacular success in the pursuit of pre-defined and sequential solutions, and of economic and other growth targets. Frequently, the conclusion is drawn that a change to pragmatic organisation is imperative, wherever instrumental set-ups still domi-nate: setting up Applied Science Foundations and mission-oriented agencies, restructuring European industries and universities, etc. It would be fatal to follow this illusory promise of quick success, instead of aiming straight at new forms of adaptive institutions. It is already too late for detours.

Adaptive organisations are geared to the flexible process of continuous search and modification which is the essence of planning at the higher levels of 'futures-creative' planning, namely the levels of strategies and policies. They may include pragmatically organised departments, or adaptive inter-organisational frameworks may be composed of 'building blocks' which are, in themselves, pragmatic organisations. Tactical plann-ing and action, aiming at well-defined and unambiguous targets, will still benefit from a pragmatic approach, which, however, is embedded now in planning and decision making processes dealing flexibly with a multitude of strategic and policy options. Adaptive organisations, or inter-organisa-tional structures, permit the systematic consideration of high-level objec-tives which may be translated into a variety of tactical targets (e.g. specific technologies, products, processes, or services) from which an operational target is selected and 'frozen in' only at a relatively late stage, imme-diately preceding action. Long-range forecasting and planning over a time scale of several decades can be practised in the fullest sense only within the framework of adaptive institutions. Here alone, the full mechanism of 'futures-creative' planning can unfold in the interaction between policy, strategic, and tactical planning.

Adaptive organisations, in considering strategic and policy alternatives, are conducive to non-linear planning and action. In particular, they will select from a wide spectrum of feasible technologies by considering their outcomes in a systems context, for example a functional framework, and investigate the possibility of simultaneously satisfying a number of systems criteria. Thus, adaptive organisations have to be developed especially for integrative planning of technology.[6] It becomes immediately clear that this complex task can often be undertaken effectively in multi-faceted inter-institutional frameworks rather than in single institutions. However, such inter-institutional frameworks must have a capacity for synthesis – in other words, a capacity for strategic and policy planning. It may be noted, that the intellectual separation of planning and decision making (or of

[6] See Chapter 3 of this volume.

planning and action) may find its expression here in a visible geographical separation. Given the dimensions of the environment against which policies of institutions such as industry, research institutes, government departments, etc., have to be planned – frequently a global environment, or functional complexes of national or world scope – it becomes obvious that common backgrounds for policies will be required in many cases for inter-institutional combinations. Strategic planning, on the other hand, will be more an affair of individual adaptive organisations.

The characteristic structure of adaptive organisations aims at a combination of powerful vertical impetus and subtle horizontal synthesis. Horizontal staff groups and planning units assume special importance. Their task requires a highly dynamic and flexible approach and may be compared to 'translating' implications of objectives as well as feasible options vertically and horizontally in all directions. They are supposed to 'animate', to stimulate creativity throughout the entire structure, to collect and homogenise inputs, and to synthesise them into strategic and policy options. In short, they take care of one of the basic requirements derived above: the combination of decentralised initiative with centralised synthesis. Furthermore, the structure of adaptive organisations is geared to dynamic programme management under stable or only slowly changing functional headings. Adaptive organisation is conducive to the introduction of the 'profit-centre' concept.[7]

The evolutionary trend toward adaptive organisation is already visible in all areas where the incentive to establish long-range forecasting and planning has been recognised. The Planning-Programming-Budgeting System (PPBS) has already become a successful planning framework for dynamic programmes, subsumed under functional categories. After the experience with PPBS in the U.S. Department of Defense, the U.S. Government has initiated, in 1965, its introduction to the civilian branches of government. Whereas this would impart an adaptive character to the U.S. Government as a whole, the scope of traditional and originally instrumentally organised government departments – such as foreign policy, defence, commerce, agriculture, etc. – will inevitably resist programme formulation in the proper functional categories. However, it can be noted that some of the newly established government departments – in the United States, for example, the Department of Transportation and the Department of Housing and Urban Development, and, in a way, also the planned merger of the Commerce and Labor Departments – are practically congruent with the functional planning categories. PPBS has also been successfully tried out at lower jurisdictional levels, such as states, counties and communities, and is also considered for international organisations.

[7] Jay W. Forrester, 'A New Corporate Design', in E. Jantsch (ed.), *Perspectives of Planning*, OECD, Paris 1969.

The best-organised industrial companies, working in areas of rapid technological change, are already[8] developing structures and management schemes leading to a degree of intra-organisational adaptivity, such as flexible 'innovation emphasis structures', superimposed over the more or less rigid administrative and operational structures. PPBS is also considered for application in industrial contexts. However, the development of inter-organisational industrial frameworks for tasks requiring an adaptive approach, is as yet barely contemplated. Where it has been realised partly, is in the framework of complex government programmes, such as the American space programme under the leadership of NASA. There can be no doubt, however, that industry will have to develop the capability to plan, design, build, and perhaps even operate the big systems of the future – systems of transportation, communication, education, health, urban living, etc. – and that this challenge can only be met by a flexible inter-institutional response from industry as well as other planning, research, and operating institutions.

In higher education, finally, the present trend toward inter-disciplinary pragmatic approaches, reflecting in present university structures, will have to be pushed much further to functional, i.e. outcome-oriented, structures developing in congruence with the functions inherent in the joint systems of technology with Nature, man, and society, and other 'bi-polar' sub-systems. In short, there will be a need for university departments such as 'Environmental Health', 'Environmental Control (pollution, etc.)', 'Urban Development' (in the integral meaning implied in Doxiadis' term 'ekistics', comprising the elements Nature, man, society, shell, networks), 'Integral Transportation Systems', etc. Above all, there is need for a university structure capable of teaching the planning, design, building, and operation of systems in an integral way. Instead of being taught scientific or technological disciplines, or special skills, the student of the future ought to learn how to build good, 'livable' systems, how to use a wide spectrum of scientific and technological possibilities to improve the quality of life and how to cope with the requirements integrative planning imposes on systems development. The primary aim of education will be to produce systems engineers in the broadest meaning of the word. They ought, for example, to deal integrally with the whole chain from product development to waste disposal or recycling of the waste. Today, the students of systems engineering may learn how to build complex techno-logical systems, such as weapons systems – a very narrow-minded approach compared with the criteria to be satisfied in viable systems, such as joint

---

[8] A. D. Chandler, Jr., in his book *Strategy and Structure* (Doubleday, Garden City, New Jersey 1966), places the beginning of the movement of American industry from pragmatic toward adaptive organisational forms into the 1920s, and its peak into the 1950s and 1960s. However, it should be noted that, whereas Chandler's Types I and II correspond precisely to instrumental and pragmatic organisational forms, his industrial Type III represents only a first step towards an adaptive organisation.

systems of society and technology. The task is not only the frequently cited marriage of the natural and social sciences, but the reformulation of the entire university curriculum in the light of the systems sciences.[9]

It may be noted here that H. Marcuse's now famous 'one-dimensional man' would only be free to move between the two extremes of being organised in a society with pragmatic institutions – the 'repressive' civilisation enforcing the principle of individual efficiency in the service of social goals, which overshadow individual goals – and a 'non-repressive', essentially anarchist society which would dispense with institutions (except those providing the material needs of man). In this view, the one-dimensional spectrum of possibilities would range from the supremacy of society and civilisation (and the 'repression' of individual goals) to the supremacy of the individual (and, consequently, the 'repression' of social goals and the concerns of mankind as a whole). If, at the one extreme, the synergistic effect of society and civilisation is led *ad absurdum* by lack of creative inputs, the negation of the psycho-social evolution of mankind and its reduction to the sum of individual creative acts (which are seen as the expression of man's *eros*), which we find at the other extreme, leads straight into the abyss of uncontrolled development and the catastrophes which can be readily forecast – in other words, to complete loss of human freedom.

Adaptive institutions, along with the integrative, normative, and adaptive character of 'futures-creative' planning, provide a genuine alternative to this one-dimensional dilemma – one may say, they add a new dimension to the application of human creativity and freedom.

PROBLEMS OF INTRA-INSTITUTIONAL ADAPTIVITY

In considering the creation of adaptive organisational frameworks, at least two basic steps may be distinguished:

(a) The superposition of a flexible 'innovation emphasis structure' (as one of the leading American electronics companies calls its innovation planning scheme) over a more rigid administrative and operations structure; and

(b) Bringing the planning and operations structures to full congruence, or even identity.

There are good reasons for believing that for many tasks of techno-logical and social innovation, taking just the first step and stopping short of the second, will be the better solution for quite some time to come. Today, the *avant-garde* of planning at both the national and the industrial

---

[9] See also Chapters 15 and 16 of this volume.

levels is generally in the process of taking or consolidating step (a). The function-oriented PPBS is superimposed over the traditional service structure Army–Navy (plus Marine Corps)–Air Force within the U.S. Department of Defense[10] and over the mainly traditional structure of departments and agencies of the U.S. civilian government.

At the level of technology step (b) – congruence of planning and operations – may even become impossible, because the results of proper planning for strategic decision making and for policies will require the utmost flexibility in the technological solutions selected for implementation. To reach this flexibility by continuously changing factory equipment and the tasks of design departments, is generally inconceivable at least at the stage of automation attained at present and attainable during the coming decade. However, the necessary degree of flexibility can be attained by changing combinations of resources 'modules' – equipment, special skills, manpower, etc. – within a company, as it can be attained through inter-organisational combinations.

It has already been pointed out in Chapter 2, that there are two basic approaches to organising for flexibility: gaining flexibility by acting from the outside or from the inside of existing structures. Typical structural solutions for both approaches are sketched in *Figure 9.1*.

Gaining flexibility by acting from the outside of an organisational structure may mean, for example, the following: An organisation form, geared to operational planning and action only, is given by a rigid structure of rigid government departments, industrial product lines, or university departments; control is from the top of each vertical column, no flexibility is aimed at. An organisation form, geared to strategic planning and action, may be composed of the same rigid vertical columns, but with horizontal staff groups charged with the development of alternative elements and with strategic synthesis, which may subsequently be used by the top executive to introduce changes in the rigid vertical columns, or to set up new vertical columns. Corporate staff groups may be mentioned as example. An organisation form, geared to policy-planning and action, would be built in an analogous way but with strategic vertical units (and more or less rigid operational 'modules' as sub-units) taking care of the strategic flexibility and with horizontal policy staff groups investigating the roles of these units and their combination. Examples for such strategic units are given in industry by function-oriented (not technology- or product-oriented) structures. The emphasis is on systemic functions (e.g. functions of technology in social systems). In universities, experiments mainly in undergraduate education go to structures focussing on such systemic functions (e.g. Colleges for Environmental Design) in some

---

[10] However, the courageous attempt of the Canadian Armed Forces should be noted, which, with the introduction of a function-oriented planning structure in 1967, abolished the traditional service structure altogether.

(a)   Organisation form geared to operational action only.

(b)   Organisation forms geared to strategic action.

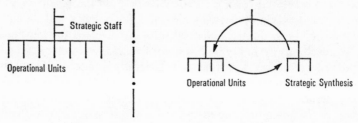

(c)   Organisation forms geared to policy action.

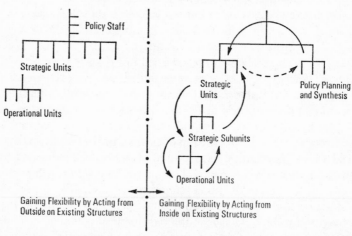

Fig 9.1 The development of organisation forms geared to rational creative action (Left-hand side: acting for systems from positions in metasystems; right-hand side: acting within systems by going to next higher level of abstraction.)

American universities. Normally, the step to establish policy-planning staffs has not yet been taken even where such strategic units have been established.

Gaining flexibility by acting from the inside of existing structures is an approach which holds perhaps more promise and is less limited. An organisation form, geared to strategic planning and action (and neglecting policy-planning) would then split between responsibilities for 'the present' and responsibilities for 'the future'. Whereas the present is taken care of by operating companies or divisions – only industry so far has gone to such organisation forms – the development of strategic elements (also through corporate level R & D) and the strategic synthesis would be part of a separate 'Corporate Development' structure. To make strategies effective in the operating units, 'innovation emphasis structures' (which are not organisational, but intellectively constructed relevance structures) are used to structure flexible task force approaches in the operating units. Decisive is, of course, a strong personality holding the two parts of the corporation together and making them interact fruitfully. No governmental structures of this type have yet been developed. In universities, inter-disciplinary centres may have a weak effect in this direction with people getting stimulated by their interaction with such centres and translating this stimulus to their work in their assigned positions. An organisation form, geared to policy-planning, becomes possible if the units responsible for operations are organised as strategic units in the same sense as above, and the high-level split is now between a group of strategic units and of policy-planning and synthesis units (which may also, to some extent, become involved in the strategic synthesis). Some examples in American industry are emerging which show such characteristics.

In considering adaptive institutions at the national level, one may believe in the development of a closer analogy between governmental and industrial structures. This analogy partly already holds for the present situation and, in the United States, corresponds to a conscious White House concept. For the purposes of full strategic and policy-planning, it may be expected to develop further to a stage depicted in *Figure 9.2*, in which a central agency for national and international 'Corporate Development' will have the task of preparing technological and non-technological (e.g. social) innovation and carrying out research and development through function-oriented agencies and laboratories – analogous to an industrial 'Corporate Development' structure. The principal difference would be that, in the government scheme, the applied research laboratories also would preferably fall under 'Corporate Development' as long as the government departments and operating agencies do not yet represent a flexible function-oriented framework (as it can be established more readily for industrial operating divisions).

The scheme in *Figure 9.2* also implies a drastic change in the funding of governmental research and development. Not only will instrumental agencies, such as the National Science Foundation and the National Institutes of Health in the United States, no longer have a place in function-oriented planning and implementation, but also funding of research and development will no more go through the government departments – representing a rigid and only partly function-oriented structure that will deal with the task of 'planning for society' in a fractionated way only – but through 'Corporate Development' to which resources would be allocated directly by Congress and the President (to take the American governmental structure). The planning groups under the same roof, concentrating on synthesising technological and social aspects, may resemble the National Goals Research Staff, founded in 1969, and floundering soon after publishing its first report.[11] Fundamental research will also come under 'Corporate Development' and be guided in a much more flexible

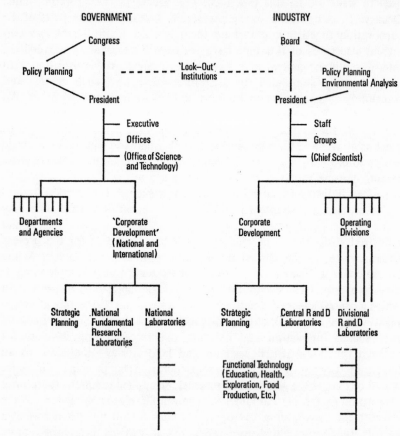

Fig 9.2   The developing analogy between structures for innovation in government and industry, demonstrated on a hypothetical example for the United States.

way, emphasising its relevance to national and international policies as well as to strategies and functions; fundamental research maturing to a stage where it contributes to functions ('oceanography' to the 'food production' and 'exploration' functions, etc.) will be moved on to functional research.

To the extent to which government will not be able to match industrial planning and implementation, for example due to the failure to establish

[11] *Toward Balanced Growth: Quantity with Quality,* Report of the National Goals Research Staff, Washington, D.C., 4 July 1970.

an adequate mechanism for innovation, government's role in 'shaping the future' will be inferior to industry's role.

Whereas the organisational structures for innovation emphasis, as they are emerging now in industry, will be adopted by government, industrial structures will evolve from the authoritarian towards the constitutional form. One of the forthcoming changes pointed out by J. W. Forrester[12] is already discernible in outstanding companies, especially in the United States: namely the 'modification of the primary objectives of the corporation away from the already diluted idea of existence primarily for profit to the stockholders and toward the concept of a society primarily devoted to the interests of its participants'. The impact of this evolution will be felt most dramatically on the technical and management levels (Galbraith's 'technostructure'). The consequences in terms of the abolishment of the superior/subordinate relationship and the establishment of the 'profit-centre' concept at the level of the individual, elaborated by Forrester, will mark the attainment of high intra-institutional adaptivity.

Even higher degrees of 'fluidity' at the level of the individual can be expected with advances in information technology and automation that are feasible before the end of the century:[13] self-adaptive inventory, production and organisation control; automated marketing through home-terminals; inter-linked data banks; automation of office and institutional data handling; regional and global education centres accessible through home-terminals; automation of optimal benefit resource application decisions; etc. The entrepreneur of the future, i.e. the individual with technical and management skill, may very well interact – through 'profit centres' – not only with one organisation, but with a multitude of organisations to which he is connected by means of his home-terminal. Thus, the pursuit of intra-institutional adaptivity leads to the dissolution of institutional boundaries and the ultimate consequences of adaptive 'inter-institutional' combinations at the level of the individual. This was to be expected if we insist on decentralised initiative and on bringing out the creative energies at the level of the individual. Man constitutes the 'atom' of institutional creativity.

PROBLEMS OF INTER-INSTITUTIONAL ADAPTIVITY

In restricting our considerations to the next decade, we may assume a certain continuity of fixed institutions – even if the form of their engagement changes and becomes more flexible. In the more distant future, some

---

[12] Jay W. Forrester, 'A New Corporate Design', in E. Jantsch (ed.), *Perspectives of Planning*, OECD, Paris 1969.
[13] Hasan Ozbekhan, 'The Future of Automation', *Science Journal* (London), October 1967.

of these institutions may lose their identity completely or may give way to entirely different institutions.

The only new institution whose creation appears inevitable if policy planning is to be established at a national and international level, is the 'look-out' institution (H. Ozbekhan), or the 'Institute for the Future' (O. Helmer).[14] In Ozbekhan's words,[15] the main function of a 'look-out' institution will be 'to conceive of possible futures; to create standards of comparison between possible futures; to define ways for getting at such possible futures by means of the physical, human, intellectual and political resources that the current situation permits to estimate'.

Figure 9.3 attempts to outline a possible 'innovation emphasis structure' at national level, as it may perhaps be expected to emerge in the 1970s, at least in the United States. It is not to be confused with a structure for financing and implementing the results of planning.

The outstanding feature of the 1970s will probably be the emergence of industry as a 'planner for society' on an equal level with government. This will result partly from industry's leadership in planning and management skills, partly from the particularly close relationship between industry and society. The consumer activities of society are satisfied directly by the industry-operated process activities – which give industry also the most favourable starting point to influence society's attitudes. It is significant that in the United States, in contrast to Europe, almost the full spectrum of connective activities – with the exception of road building – is also planned and operated by the process and service industry. If the private sector is responsible for the dramatic advances in the communications networks, and sets out to penetrate into city building, there is no reason to exclude it from regional development.[16]

Recent global estimates within a twenty-year time-frame foresee that the current process of industrial concentration will result in approximately 600 or 700 industrial groups, most of them operating as trans-national companies, which will act as prime contractors for complex technological developments and sub-contract to a large number of small specialised firms. These 600 or 700 industrial groups can be expected to form the creative basis not only for providing technological contributions to social functions, but also for the planning, design, building and operation of joint

[14] The first 'Institute for the Future' in the meaning given by O. Helmer, started at Middletown, Connecticut, in the fall of 1968 under Helmer's direction.

[15] Hasan Ozbekhan, *The Idea of a 'Look-Out' Institution*, System Development Corporation, Santa Monica, California, March 1965.

[16] It should be noted that, with urban development having already entered the stages of 'metropolis' and 'megalopolis' (e.g. in the American North-East corridor, or the Great Lakes region), a vacuum between community and national administration is felt particularly in the development of networks. On the other hand, American industrial companies have already become active in economic and general development planning on a national and regional level (e.g. Algeria, Sudan, Crete, and the West Peloponnese, etc.). See also Chapter 13 of this volume.

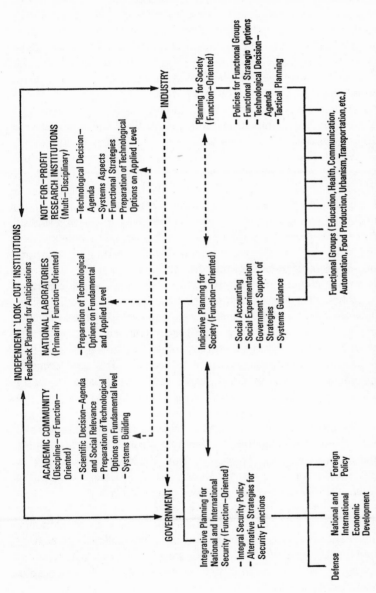

Fig 9.3 A Possible 'Innovation Emphasis Structure' at National Level. Only the principal lines of interaction are indicated. (This 'emphasis structure' must not be confused with structures for financing and operating R&D programs, which will look quite different.)

systems between technology and man, or Nature. They will participate in, or lead, flexible and adaptive inter-industry structures – consortia, joint ventures, prime-contractor/sub-contractor patterns, flexible groupings of industry together with national laboratories and not-for-profit research institutions, and flexibly engaged discipline-oriented organisations (e.g. in academic and other fundamental research). The success, and added stimulation, reached by the engagement of some 20,000 or more industrial companies for NASA's projects in the United States, already seem to prove the feasibility and the value of such an approach.

The overall national planning scheme, as suggested in *Figure 9.3*, rests on the two 'pillars', government and industry, both of which concentrate their planning on those areas where their power of subsequent implementation is greatest – government on national and international security and industry on the joint systems which society forms with technology. Governmental planning for security, currently done piecemeal, should become an integrated exercise between, if we take the United States Government as an example, the White House/Department of Defense/State Department triangle, with inputs from departments and agencies concerned with economic development and aid, trade, and monetary discipline. Such integrated planning for security (including numerous cross-links with 'planning for society') will become even more important when the global crisis due to the population explosion, the inability of the agricultural food production system to cope with the rising food demand, the widening 'gap' between the rich and the poor countries, etc., will reach much larger dimensions. The country that knows only how to defend itself but not how to manage the crisis (with technological developments providing effective means in many ways) and possibly avert a major catastrophe, may well be doomed before the end of the century.

The government's contribution to 'planning for society' will be primarily of an indicative nature and concentrate on the augmentation and unification of industrial planning, for example through social accounting and social experimentation. But government will have a crucial role in bringing the plans for large systems to fruition, for example by creating new markets such as the recently proposed 'pollution market'. Such new markets have been created in the past, for example, by the U.S. Atomic Energy Commission (before civilian nuclear energy became an economic proposition) and by NASA. The planning and implementation of technological and social change through large systems will create new relationships between government and industry and change the modes and criteria of the economic system thoroughly.

In spite of the very close interaction to be expected between government's and industry's shares of planning, there must be an institution capable of integrating both and of thinking in terms of full-size anticipations (possible futures): a *'look-out' institution*, possibly financed through government and industry, but acting independently.[17] Its role in the scheme, according to *Figure 9.3*, would not only be to construct and evaluate anticipations by putting together the planning pieces received from government and industry, but – much more important – to stimulate and orient planning by providing an 'over-view', and to ask the right questions by sub-contracting tentative planning tasks primarily concerned with alternative ways (strategies) to achieve specific anticipations. The powerful 'alignment' of thinking through self-motivation which is experienced today wherever strategies and policies are formulated clearly (especially in industry), can be expected to repeat itself at a higher level if anticipations, and the ways to get there, can be spelled out clearly. However, the more difficult task of the 'look-out' institution will be to keep the pattern of anticipations in a 'fluid' state by operating a continuous intellectual feedback process. To a certain extent, universities may come into a position to play the roles of 'look-out' institutions.[18]

It is difficult to imagine today, in a situation characterised by a strong polarisation of all national systems aspects toward government and, to a lesser degree, toward industry, that the spiritual leadership in planning will be placed in a 'brokerage institution' without executive power. Nevertheless, this appears to be a better solution than burdening the government with planning tasks beyond its horizon, and requiring an intellectual flexibility that cannot be maintained in close contact with decision makers. Strategic planning by consulting 'think groups', exemplified in the United States by the RAND Corporation, has reportedly become a decisive success in defence planning. The recent decision, in the United States, to engage RAND, RAC (Research Analysis Corporation) and other consulting groups in the civilian area in connection with the introduction of the PPBS (Planning-Programming-Budgeting System), can be interpreted as a first step in the direction of establishing a planning scheme such as discussed here.

Another bold, but logical feature of the scheme according to *Figure 9.3* is the role of planning of the academic community (Academies of Science, universities, etc.), the national laboratories, and the not-for-profit research institutes: in planning, they are primarily interacting with the 'look-out' institution, not with their financial sources government and industry. The establishment of a governmental 'Corporate Development' function, as shown in *Figure 9.2*, will facilitate the realisation of such a concept, in particular in cutting the ties between national laboratories and government departments and ministries.

In the United States, the academic community has made a first step in the direction of participating in planning at national level, through the COSPUP (Committee for Science and Public Policy) of the National

[17] Such a 'look-out' institution has been proposed in Sweden.
[18] See Chapter 16 of this volume.

Academy of Sciences[19]– and even for this very first step, there is no 'climate of opinon' yet to be found in Europe.

It would be highly desirable if national schemes aiming at 'planning for society' could soon be supplemented by *international schemes*. The existing international organisms are too inflexible to become part of an 'innovation structure'. However, they might become instrumental in setting up suitable groups, such as 'international look-out institutions of the advanced countries' which would first concentrate on planning for problems that can be solved only on a world-wide basis,[20] or regional 'look-out' institutions.

It has been observed above that the development of the art of planning is pushing the development of institutions at present. The focus on systems planning – especially the integrative planning of joint systems of technology and society – will eventually change and mould institutions so as to give them the capability to gain an over-view over the system and synthesise possible ways of conceiving, building, and operating it. Clearly, the geographical and other systems boundaries frequently do not coincide with the institutional boundaries which our political systems recognise today: economic and defence blocs, nations, states, counties, and communities. The call for integral approaches to technological development on a European basis bears testimony to the dominating exigencies of technological change, which already overshadow traditional considerations of economic and defence agreements.

Political blocs, economic and defence blocs, nations, and communities may be regarded as pragmatic institutions pushing single-track concepts. These pragmatic institutions will be gradually absorbed by, or subordinated to, new adaptive institutions which will take forms that will enable them to cope with the future systems aspects of our planet, especially with the joint systems between technology and society, and technology and Nature. Such new adaptive institutional frameworks, transgressing national and other traditional boundaries, will finally permit the overall problem of 'ecological engineering'[21] to be tackled on regional and global bases.

[19] See Chapter 7 of this volume.
[20] This has been proposed in: Aurelio Peccei, *The Chasm Ahead*, Macmillan, New York 1969.
[21] See Chapter 3 of this volume.

10. Organisational Structure and Long-Range Forecasting

THE EXPANDING SCOPE OF FORECASTING

FORECASTING takes a look ahead at potential dynamic developments. There are obvious inter-dependencies among such developments in different domains – economic, social, political, technological, and psychological, for example. Thus, the integrative approach,[1] emphasising concern with the dynamics of multi-dimensional systems, is becoming increasingly accepted as a conceptual and methodological basis for long-range forecasting and planning. Both the long-range view and the hierarchical structure of normative forecasting and planning enforce a dynamic approach embracing many dimensions of change.

The inherent difficulties of integrative forecasting become apparent in the naive attempts to obtain a synopsis by combining forecasts obtained more or less independently and along one-dimensional lines of development. The objective is frequently to derive a scenario, a cross-section at some given moment in the future. Most technological forecasts are still derived individually from relatively autonomous criteria and subsequently merged with the other ingredients of a dynamic scenario.[2]

This would have been more appropriate for the recent past when the thrust of technological development was less guided by social, ecological, psychological, and anthropological criteria than it probably will be in the future.

If we reverse – as, indeed, we should – the currently dominating principle of 'adapting society to technological innovation', to make it 'adapting technology to the evolution of society', we make technological forecasting more difficult in two ways: First, forecasted technological

---

[1] Frank P. Davidson, 'Macro-Engineering – A Capability in Search of a Methodology', *Futures*, Vol. 1, No. 2, December 1968. See also Chapter 3 of this volume.

[2] Herman Kahn and Anthony Wiener, *Toward the Year 2000: A Framework for Speculation*, Macmillan, New York 1967. It is also astonishing to see the authors extrapolate up to the year 2000 the concepts of power and economic strength in the rigid framework of a multiplicity of nation-states as they are defined today.

feasibilities will mean little in themselves if the normative elements are to be derived primarily from social criteria – we will need fully developed social forecasts against which we then may match these technical feasibilities. And, second, the entire institutional framework, as it exists within society, must itself be taken as a variable in any social forecast.

This implies that forecasting for the purposes of a specific instrumentality – an industry, a corporation, a government agency, etc. – can no longer assume that instrumentality and the institution of which it is part to be static. The forecast must include the evolution of the institution itself: its varying structures; its evolving pattern of resources; its dynamic inter-relationships with other existing, emerging, or decaying institutions; and the adaptation of its responses to different sets of challenges. Technological forecasting, geared as it is to the purposes of institutions acting as agents in technological change, is to a particularly high extent confronted with this problem of institutional change; for technological change takes place at a very high rate and is very closely coupled to social change. From the above it follows that useful technological forecasts must have two attributes:

1. Forecasts made for a specific institution must be fully integrated with forecasts for the entire relevant social system, beyond the specific environment with which the institution directly interacts.
2. Forecasts must emphasise the potential for continuous self-renewal of the institution for which they are made.

Looking at technology, and the way it interacts with society, we may distinguish three levels of organisational complexity. These may also be understood as levels of hierarchical dependence:

'Joint systems' of society and technology (e.g., systems of urban living' communication/transportation, health, food production, etc.);[3]
Missions of technology within society (e.g., shelter, personal transportation, cure of diseases, agriculture, etc.);
Specific technologies (e.g., prefabricated construction elements, internal combustion engine cars, specific drugs, hybrid plant species, etc.)

Today, forecasting activity focusses primarily on the lowest level, and many of its techniques encourage staying there. In industry, this type of forecasting favours the development of product lines and specific processes to their ultimate possibilities. It also conveys the comforting – but misleading – feeling that institutions will not have to change.

Some organisations – government agencies as well as industrial corporations – are at present more or less successfully trying to come to grips with the complex forecasting problems at the middle level. The United

[3] For further elaboration of the concept of 'joint systems', see Chapter 3 of this volume.

States Government's Planning-Programming-Budgeting System (PPBS) is designed to encourage and facilitate forecasting and planning at this level, frequently referred to as the 'strategic' level. In industry, the introduction of corporate long-range planning has gradually led some corporations to reformulate their goals in terms of broad missions, or functions of technology *vis-à-vis* society.[4] Here, the institution which gives rise to technological change, accepts for itself a dynamically changing role, while retaining its basic identity.

Today most institutions hesitate to acknowledge any responsibility for integrating their own pattern of actions into the wider concepts of the 'joint systems' constituted by society and technology. However, this issue will soon become a matter of life or death for the existing institutions, especially for industry. If companies choose to continue acting as independent entities pursuing the goals inherent in their own (present) structure, and interacting with other independent units, new institutions will be forced to assume responsibility. Instead of industry, giant publicly-owned monopolistic enterprises of regional or world scope may take over the planning, and possibly also the production and distribution tasks in entire areas defined by social needs. However, the existing institutions are still free to participate actively in the present move towards a more fully integrated society. They can even assume a leading role in this movement.[5]

The task of technological, more appropriately, integrative, forecasting, for any institution may be divided in two. The two subtasks determine the ways one may properly organise the forecasting function:

—Forecasting the role of the institution in a broad societal context beyond the environment with which the institution interacts directly.
—Forecasting the deployment of the resources available to the institution to fulfil the forecasted roles.

The types of organisation suited to these two subtasks will be briefly discussed in the following sections.

First, a short note on technological forecasting methodology: Most exploratory techniques in use at present (perhaps with the exception of envelope curve techniques) tend to enhance a static and rigid position of the institution. They facilitate linear and sequential modes of forecasting and planning. On the other side, normative techniques, particularly relevance tree techniques, contribute significantly to the clarification of alternative technological options falling under specific missions. These techniques thus favour a flexible and adaptive position of the institution itself.

As Roberts points out, exploratory techniques are about to enter a stage of development comparable to complex multivariate econometric

---

[4] See Chapters 4 and 8 of this volume.
[5] See Chapter 13 of this volume.

forecasting in the economic area.[6] In the process they will climb one step in their usefulness for integrative forecasting. Concurrently, feedback models linking exploratory and normative aspects may soon come into the stage of large-scale application. It is evident that forecasting institutional change will depend a great deal on feedback thinking and simulation.

FORECASTING THE ROLES OF INSTITUTIONS

The trend away from the present fragmented social systems – corporations, cities, nations – and towards a more fully integrated system of world society is inherent in the psycho-social evolution of mankind itself. We are standing at a new 'threshold' of this evolution, comparable in importance to those thresholds crossed by mankind when it developed primitive agriculture or trade links between villages and cities.

The unity of mankind and the indivisible responsibility for the planet Earth constitute the challenge at this new threshold, which must be crossed more rapidly than was ever required before. The response has to be formulated and implemented in the decades left of this century. If it turns out to be unsatisfactory, not only will mankind fail to achieve the necessary degree of integration, but severe dynamic instabilities will throw us back to lower levels and earlier phases of evolution.[7]

For many problems at the macro-level of the 'joint systems' formed by society and technology – food, environment, development, etc. – nothing short of such a global view will be required in the horizontal integration of forecasting and planning.

To industry, the integration of its processing activities will become of particular concern. Product and service lines, developed – and frequently pushed to their extremes – in a piecemeal way, interact within larger systems, such as the systems of urban living and of the human environment. Linear and sequential modes of action – in other words, the indiscriminate pursuit of feasibilities in a fragmented way – inhibit the healthy development of such systems and the co-ordination of the *outcomes* of lines of action.

Inevitably, industry will face the paramount task of inventing, planning, designing, building, and possibly also operating, such joint systems of society and technology. Inherent in technological innovation is the task of social innovation. Technology must be forecasted and planned in this wider connotation.

[6] Edward B. Roberts, 'Exploratory and Normative Technological Forecasting: A Critical Appraisal', *Technological Forecasting*, Vol. 1, No. 2, November 1969.

[7] For the most comprehensive statement on this issue, see the following book written by an industrialist, Aurelio Peccei, *The Chasm Ahead*, Macmillan, New York 1969, as well as R. Buckminster Fuller, *Operating Manual for Spaceship Earth*, Carbondale, Illinois 1969.

This implies that forecasting and planning also have to be vertically integrated. They become part of inventing, planning, and creating the future of society, not just of the particular institution. Thus far, implementation and goal-setting have been pursued separately, the former by independent institutions such as industry, and the latter by government. But now integration of the goal-setting and implementing activities is becoming necessary to deal appropriately with the 'joint systems' of society and technology.

Competition will increasingly assume the form of competition between ideas and plans, inventions and designs of systems – in other words, between social software alternatives rather than industrial hardware. This has led some to anticipate a shift in the roles of industry and university. Daniel Bell,[8] for example, writes: 'Perhaps it is not too much to say that if the business firm was the key institution of the past hundred years, because of its role in organising production for the mass creation of products, the university will become the central institution of the next hundred years because of its role as the new source of innovation and knowledge.'

There may be a fallacy here in assuming that industry will not change in its institutional character with this new challenge. Historically, industry has taken over from the university the bulk of applied scientific research and technical development and is now becoming deeply involved even in fundamental research. There is no reason why industry should not acquire the capability to deal imaginatively with the new challenge presented by the 'joint systems' of society and technology. However, one may agree more readily with Daniel Bell when he continues:[9] 'To say that the major institutions of the new society will be intellectual is to say that production and business decisions will be subordinated to, or will derive from, other forces in society; that the crucial decisions regarding the growth of the economy and its balance will come from government. . . .'

The pressures emanating from this development may well lead to the emergence of three of the existing types of institutions as 'joint planners for society': government (which may gradually change as the principle of the nation-state increasingly gives way), industry, and the university. Both industry and the university ought to become *political institutions* in the broadest sense – contributing to an integrated planning process for society at large, instead of pursuing independently conceived subgoals.[10]

The central problem here may be seen in analogy to the problem of organising forecasting and planning within a corporation, as will be discussed below. But the degree of complexity, and therefore the danger of

[8] Daniel Bell, 'Notes on the Post-Industrial Society', *The Public Interest*, No. 6 (Winter 1967).
[9] *Ibid.*
[10] Erich Jantsch, 'Technological Forecasting for Planning and Institutional Implications, in Proceedings of the symposium on *National R & D for the 1970's*, National Security Industrial Association, Washington, D.C., 1968.

failure, in getting this subtly integrated process well under way is much higher at the level of society. But, in analogy to the corporation, the principle of *decentralised initiative and centralised synthesis* provides the only effective way of soliciting creative contributions from all parts of society. The outstanding role of industry and the university merely derives from their effectiveness in organising the intellectual capacity of society. If industry should prefer to stagnate and concentrate on optimising mass production techniques, think tanks in some form may emerge as competitors with ideas and plans. Industry could become an appendage.

At present we may assume that both industry and the university – after a period of resisting change and clinging to the old principle inevitably evoked in pertinent discussions today – will accept their new roles as political institutions and develop systems laboratories focussing on the 'joint systems' of society and technology. These systems laboratories would then provide the organisational framework for the bulk of technological forecasting. Technological feasibilities would be tested in the context of the dynamic behaviour of complex social systems. This would finally result in the purposeful pursuit of technological options. One would build new systems structures and no longer simply adapt social systems to new technologies.[11]

While the university may then compete with industry on equal terms in selling its ideas and systems designs, it may also be called upon by government to provide prototypes by developing experimental designs in strategically selected sectors. It may study, or contribute to the study of, systems too large to be dealt with by individual corporations. And it may organise and back up the study of global problems which will be so crucial for the future of mankind.

For a transitory period, industry may utilise to an increasing extent the services of non-profit research institutes which will offer a considerable capability in forecasting and systems planning.[12] More and more of the large industrial corporations are recognising the desirability of projecting their future roles against a common background, while retaining the principle of competition in planning and action. They recognise the necessity to harmonise their policies at the societal level, while competing with each other at the level of strategies. If, for example, a thorough background study would establish beyond doubt the necessity to develop before the end of the century non-agricultural food production technologies, and these findings would also be accepted at the political level,

[11] On the principles of purposefully designing social systems, see Jay W. Forrester, 'Planning Under the Dynamic Influences of Complex Social Systems', in *Perspectives of Planning*, OECD, Paris 1969; and Jay W. Forrester, *Urban Dynamics*, M.I.T. Press, Cambridge, Massachusetts 1969.

[12] The prototype of a 'pure' institute of that sort is the Institute for the Future in Middletown, Connecticut. Other research institutes, such as the Battelle Memorial Institute, Stanford Research Institute, and A. D. Little, have shifted from concentration on hardware research to developing capabilities in these new areas.

decisions leading to the development of alternative technological strategies would be much easier for the individual corporation.

Harmonisation of basic policies within a broad societal context would open the way to an important gradual shift in human values – easing up on the overstressed value 'competition', which governs relations between nations, institutions, corporations, and individuals today in Western civilisation and supplementing it by the value 'responsibility'.

Ultimately, a 'common policy background' may be developed in many areas of concern by concerted actions of governments – and through them of universities and research institutes – in a joint effort of the advanced countries. This may be a way to approach the problem of planning on a planetary scale and fighting the disruptive forces developing in our society and our planet.

If the mobilisation of industries and universities may be expected to ensure the creative inputs, the decentralised initiative, new types of organisations will be required to synthesise them. Synthesis will become necessary for any scale of system – communities, nations, regions, continents, and the entire planet Earth. There can be little doubt that this function of synthesis will come to dominate the form of 'systems government'; however, this will be defined, then, in similar ways as it already impresses itself on the organisation of industrial management and starts to influence the structures and functions of national government. To deal with the most pressing and threatening problems of the future, those of planetary dimensions, a World Forum has already been proposed, which would be set up jointly by the advanced countries of the world.[13] Its activities would focus on forecasting and simulating the consequences of changes in policy (systems structure) and various courses of action impinging on the present. The World Forum would also serve in the role of synthesising. It would have only a small permanent staff, but could elicit a maximum of creative inputs from all participating countries and bodies. It would involve them in a feedback loop of translating objective consequences of alternative courses of action and potential implications of recognised feasibilities into each other within a dynamic systems context.

This same principle of feedback and translation is already becoming applied in the organisation of forecasting within an institution, as will be briefly discussed in the following section.

ORGANISING FORECASTING WITHIN AN INSTITUTION

The same principles that should guide forecasting and planning in the broad framework of society – leading to a clear statement of the individual roles of interdependent institutions – also apply to the organisation of

[13] Aurelio Peccei, *op. cit.*

T.P.A.S.F.—M

forecasting and planning within an institution. Only their difficulties are reversed: The 'translation mechanism' may be established more effectively, within an institution, but it will encounter considerable difficulties in embracing and channelling through the overall systems view obtained at the level of society. The temptation to short-circuit the feedback between creative individuals working in an organisation and the concerns of society at large, will be great. Institutional, piecemeal objectives may replace societal objectives as they have replaced them in institutional planning so far. In industry, for example, this led to the crude 'maximisation of profit' motive which can no longer provide a real challenge to the creative professional. Another danger has been evoked by Galbraith with the notion that the 'technostructure' – the creative technical and mid-level management cadres – may pursue false objectives inherent in technology itself.[14] These can become a challenge to the professional if he is lacking the framework of reference to work for a real purpose.

What is needed, and more and more explicitly understood in advanced-thinking industry today, is coupling between institutional and instrumental objectives. Corporations, forecasting their role in a changing environment, will reformulate their own objectives in the light of the anticipated societal context. All forecasting and planning within the instrumentality will be aligned to this dynamic view of the institution's policy. Flexibility has to be gained by acting from inside existing structures[15] if institutional policy is to be reformulated rather than perpetuated.

The first requirement for the organisation of forecasting within an institution is therefore the establishment of an *'environmental radar' and a policy-planning forum*. Clearly, since the transformation of the institution itself is subject to forecasting and planning at this scale, this must be the job of top management and the Board of Directors. Few industrial corporations or universities have yet set up such a function. Typical organisational forms include the appointment of 'Officers of the Board', working directly for the Board and the President of a company, and high-level 'Policy Committees', meeting regularly, and perhaps supported by a small staff.

It is also the top management's task to keep the vertical translation process between objectives and opportunities running in both directions, upwards and downwards. This vertical stream of information implements the feedback process between exploratory and normative technological forecasting, involving the entire institution. In some advanced-thinking industrial corporations, Vice Presidents devote more than half of their time to this translation process. They thereby create self-motivation at all levels and elicit creative ideas which become parts of the institution's forecasting and planning process.

[14] John K. Galbraith, *The New Industrial State*, Houghton-Mifflin, Boston 1967.
[15] See Chapter 9 in this volume, in particular *Figure 9.1*.

Whenever top management fails to recognise its role in this process, the feedback between normative and exploratory forecasting will be severely curtailed. This is generally the case today. Imaginative people at lower hierarchical levels will be uncertain of the dynamic changes anticipated for the institution itself, or will have to assume it as rigid, and their inadequate attempts to link institutional and social futures will sooner or later lead to frustration. Top management sits at the top of an organisation not to control but to guide the human fabric of it. It should attempt to synthesise any responses to a challenge in form of an explicitly spelled out policy in a societal context.

The next level down is that of *forecasting alternative technological strategies to contribute to recognised technological missions* which the institution has adopted under its role. There, the 'vested interest' in specific technologies has to be overcome, as this prejudice is embodied not only in capital investment and special human skill, but also in the general thrust of specific technological development lines. People identify strongly with specific technical objectives. The foremost task at this level is to educate corporate personnel to look beyond specific projects to the missions of technology and to their implications in the context of society.

Two organisational principles help to overcome the difficulties in obtaining a perspective of the future that is not obstructed by barriers inherent in particular technologies:

—Build divisions of the institution around technological missions rather than around specific technologies or product and service lines; and

—Create a strong 'Corporate Development' function, in which alternative strategies for missions are synthesised and assessed, newly developed (e.g., in corporate-level research laboratories), and viewed in the context of dynamically evolving corporate policies; this leads to a high-level organisational split between the 'present' and the 'future' of the institution.

Several industrial corporations have adopted both principles and organised themselves so as to bring the long-range future into sharper focus. *Figure 10.1* sketches the example of Westinghouse, the first multi-billion-dollar corporation to adopt such a structure.

The same principles may also be applied to non-industrial organisations, in particular to government[16] and to the university.[17] In the future it will no longer be the functions of forecasting and planning looking for a place in the current organisation, it will be the long-range perspective on the social and technological future determining the organisation of the institution. The trend is already visible.

[16] Erich Jantsch, 'Technological Forecasting for Planning and Institutional Implications', Proceedings of the Symposium on *National R & D for the 1970's*, National Security Industrial Association, Washington, D.C., 1968. See also Chapter 9 of this volume.

[17] See Chapters 9 and 16 of this volume.

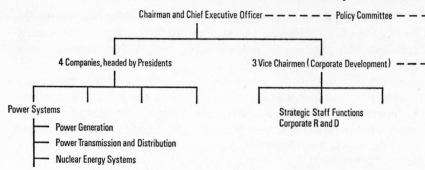

Fig 10.1 The high-level split between the 'present' (operating companies) and the 'future' (corporate development), as exemplified by the recent reorganisation of Westinghouse Electric Co. (Source: *New York Times*, 12 January 1969). Note, in particular, the structural sub-units built around missions such as 'power generation' and 'power transmission and distribution,' replacing the traditional units focusing on specific technologies, such as steam turbines, gas turbines, generators, transformers, etc.

The flexibility required by advanced corporate planning cannot be attained in current administrative and operating structures, though higher degrees of automation may change this in the future. Today, a common, satisfactory solution is to superimpose a flexible 'innovation emphasis structure' over the more rigid administrative and operating structures.

Fig 10.2 'Innovation emphasis structure' (example from an American electronics company). This structure, which is superimposed over the administrative structure, resembles a relevance tree and enforces normative thinking at all hierarchic levels.

Figure 10.2 shows such a simple 'innovation emphasis structure', worked out and maintained by the 'Corporate Development' side of an American electronics company. Ten-year strategies are updated annually through a process involving the entire company, including the operating divisions. As forecasted developments become adopted and mature, they give rise to the corresponding organisational changes. The important point is that current organisational structures are kept completely outside this framework of long-range forecasting and planning. To make this scheme work with only a very small central synthesis group for strategic planning, the principle was adopted of, eventually, having 1,000 'general managers' (denoting people at different hierarchic levels, but with equally free access to pertinent information as the company president) throughout the company.

Technological forecasting, as part of planning, depends on decentralised initiative and centralised synthesis. *Figure 10.3* proposes a scheme of involvement in this process, viewed for an industrial corporation, by successive stages of a typical technological innovation.[18] The responsibility for synthesising technological forecasting may thus be shared as follows:

—Technological forecasting for policy making at the Board and top management level;

—Technological forecasting for strategic decision making at the corporate level, e.g., by horizontal staff groups (usually incorporating mixed scientific, technical, and marketing expertise), working for top management and specifically within the framework of corporate long-range planning;

—Technological forecasting for tactical decision making at divisional management level, also preferably by staff groups.

Figure 10.3 also demonstrates the decisive role of horizontal, corporate level staff groups in synthesising technological forecasting. For three or four of the six forecasting stages, the synthesis is best entrusted to such staff groups.[19]

The complete organisational framework of technological forecasting may thus also be viewed as unfolding by the interaction of synthesis at three different levels of an organisation. But the involvement of creative people at all hierarchical levels and the stimulation of decentralised initiative are decisive. A particularly effective way of stimulating such institution-wide involvement has been found in comprehensive joint forecasting 'rounds' involving management as well as key technical people, for example on the basis of 'Delphi'-technique panels.

This integral process involves the whole institution and reaches beyond its boundaries. With the participation in broad environmental forecasting and policy-planning it reaches beyond the boundaries of the environment with which the institution directly interacts. Through this process the organisational problems deriving from the requirements and opportunities of forecasting and planning may now be seen as becoming *identical with the problem of self-renewal of the institution.*

The emphasis on institutional roles moves more and more to ideas and plans dealing with the 'joint systems' constituted by society and technology.

[18] See also Chapter 4 of this volume, in particular *Figure 4.1.*

[19] The OECD survey found such corporate level staff groups in 20 out of 23 American and in 26 out of 39 European companies considered. However, in all but one American, and in 16 European companies, they were supplemented by forecasting groups in the research laboratories and the operating divisions. See Erich Jantsch, *Technological Forecasting in Perspective*, OECD, Paris 1967.

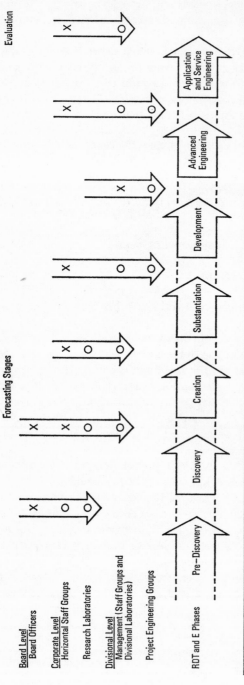

Fig 10.3 THE ORGANISATION OF TECHNOLOGICAL FORECASTING WITHIN THE CORPORATE STRUCTURE. Each forecasting stage involves the interaction between different levels. The synthesis is always made at the highest level involved. (This schematic representation does not show the full feedback taking place at each forecasting stage.)

Individual entrepreneurship within the institution will thus find increasing freedom to interact with explicitly stated policies. Instead of a split between 'Corporate Development' and operating units, a new type of 'dynamic systems laboratory' may develop – both in industry and in universities. This new organisation would form units which focus neither exclusively on the future nor on the present, but rather on dynamic behaviour of large systems. Instead of deploying in a flexible and dynamic way relatively static 'modules' of skill and equipment, as in current long-range planning of product development, emphasis on the 'software' output of institutions – on ideas, plans, and designs of systems – will permit the organisational re-unification of planning and action.

The 'dynamic systems laboratories' will then be the natural place for technological forecasting, an integral part of its activities, involving all its personnel. If, at present, tactical planning may be entrusted to the individual carrying out the action, provided that a policy is stated explicitly, the inventor and planner of systems will in the future become responsible for strategic planning, too. Since competition between organisations and institutions will take place as competiton between strategies, the principal interaction within an institution will be the dialogue between policies and strategies. With this simplification inside the institution, the way may hopefully be free to acquire a new level of planning in the interaction between institutions: the planning of society's policies.

The people animating the institutional activities may then come close to Forrester's[20] vision: 'To deal with the system of which engineering is a part and thereby to make engineering more effective within that system, the engineer must understand the system components and structure. He needs an insight into the nature of the corporation, the tasks of top management, human motivation, the relationship of the individual to the organisation, and the psychology and processes of innovation and change. The engineer could become a "change agent" to precipitate improvements in our social system. . . . He would try to clarify the enduring goals and objectives for his organisation and the people within it. . . . He would give more attention to the surrounding social system as a whole rather than as an array of isolated parts. . . . He would consider the transient and steady state behaviour of his organisation from the same system viewpoint that he approaches complex physical systems. He would strive to perfect models of social processes that permit simulation studies leading to a better understanding of organisation, information links, and policy. And he would bring to actual human organisations the courage to experiment with promising new approaches based on a foundation of design.'

Technological forecasting is a means to achieve this foundation of design.

[20] Jay W. Forrester, 'Common Foundations Underlying Engineering and Management', *IEEE Spectrum*, Vol. 1, No. 9, pp. 66–77 (September 1964).

11. From Stratified Thinking to Multiechelon Co-ordination

THIS chapter reviews a book[1] which is of great importance for the emerging new theories of forecasting, planning, decision making, organisation and of building and changing the institutions of society. It has been eagerly awaited by all those who have tried to follow the scattered publications reporting over the past decade on the work of the Systems Research Center of Case Western Reserve University in Cleveland, and particularly of its director Mihajlo Mesarović. The results of this work constitute an entirely new theory – new both in scope and content – expressed through verbal concepts as well as through rigorous mathematical treatment exhibiting a number of innovative features. What sometimes, in the succession of articles dealing with partial aspects of the new thinking, appeared as a labyrinth of abstract ideas and mathematical problems, is now presented in the book with remarkable lucidity and economy.

One cannot help feeling that, as in several previous and now famous cases, the beautiful building of mathematical theory has been constructed without too much 'normative' stimulus from the side of potential occupants – and parts of it may stand empty for some time to come. However, the authors were not merely interested in a new formalism, but tried to identify, and derived part of their motivation from needs to deal with real systems of technological, biological, or societal nature in new ways. Above all, they perceived a clear need to reformulate organisation theory for its application to complex human decision making. The first chapter of the book summarises the incentives which the authors have found in the real world, but the discussion remains somewhat vague and fails to make the strong case for the practical uses of the new theory which, in fact, is implicit in the present situation of man and his systems of living.

What, then, is the gist of the alleged relevance, and why is this book reviewed in a volume devoted to technological planning and social futures? The answer is perhaps not self-evident: One of the three types of

[1] Mihajlo D. Mesarović, D. Macko, and Y. Takahara, *Theory of Hierarchical, Multilevel, Systems*, Academic Press, New York and London 1970.

hierarchical systems discussed, the *multiechelon system* (also called 'multi-level, multigoal system' in the authors' earlier publications), is one of the principal keys to the cybernetic processes of changing values and norms, formulating objectives and policies, defining goals, and renewing institutions, which form the subject of 'futures-creative' planning and the entire process of rational creative action with a long-range and system-wide scope. In other words, the essence of the inspiring attempts to grasp the human condition and possibilities for human action in a difficult situation – as verbally conceptualised, in the first line, by Churchman,[2] Ozbekhan,[3] and Vickers[4]– is matched here by a new systems concept and its mathematical theory. More than two-thirds of the book are devoted to multi-echelon systems and the principle of *co-ordination* which makes them work. Readers who are interested merely in the basic concepts for hierarchical systems and the foundations for a mathematical formalisation of such systems, may leave aside Part II of the book which deals with the intricacies of a mathematical theory of co-ordination.

The three types of multilevel hierarchical systems treated by Mesarović and his associates, are the following ones:

1. *Stratified systems* are composed of strata (or levels of description or abstraction) each of which has its own set of concepts, principles, terms, and ways of describing the overall system's behaviour. These descriptive sets are developed in isolation and cannot be derived from each other, in spite of the hierarchical inter-dependence of the different strata (moving downward across strata increases detailed explanation, moving upward increases understanding of significance). The selection of strata depends on the observer, his knowledge and his interest in the operation of the system. Much of conventional (disciplinary) science is organised in such a stratified way (e.g., organisms, organs, cells, genes, molecules, etc.)

2. *Multilayer systems* are composed of layers (or levels) of decision complexity. They are useful, for example, in ordering the steps to be applied to sequential problem solving (a concept applicable mainly to complex technological tasks, not to human systems). In a planning context, such a multilayer system may be designed starting with the selection of a specific strategy and continuing with the sequence of steps which are part of operational planning in the pursuit of a specified, clear-cut goal.

3. *Multiechelon systems* are composed of echelons (or levels) of subsystems, some or all of which – at all levels – may be goal setting and decision units, which are co-ordinated by decision units at the next higher hierarchical level. It is important to understand that the function of the higher-placed units is not one of rigid control, but of co-ordination. In a multiechelon system, no goal is defined *a priori*, or from the top, for the overall system – it results from the interaction of subsystems goals, to the

[2] C. West Churchman, *Challenge to Reason*, McGraw-Hill, New York 1968.

[3] Hasan Ozbekhan, 'Toward a General Theory of Planning', in E. Jantsch (ed.), *Perspectives of Planning*, OECD, Paris 1969.

[4] Geoffrey Vickers, *Freedom in a Rocking Boat: Changing Values in an Unstable Society*, Allen Lane, The Penguin Press, London 1970.

extent that they can be co-ordinated, and it may become modified if goals or co-ordinability change anywhere in the system. Energies and ambitions at *all* systems levels are brought into play, and the system is 'alive' in many or all of its elements, thereby justifying the hierarchical structure. The general case of the multiechelon system is the multilevel, multigoal system; single-level, single-goal systems and single-level multigoal systems are special cases.

It should be noted that formalised approaches to human decision making have so far mainly focussed on single-level multigoal systems. This holds for the theory of games developed by Von Neumann and Morgenstern,[5] for the theory of teams (co-operative stituations) proposed by Marschak and Radner,[6] and also for the theory of complex feedback systems and their dynamic modelling, as developed and applied so far by Forrester.[7] Most of the older versions of organisation theory are static or adhere to single-goal concepts. Some of the behavioural approaches which abound in English-language literature since more than a decade, come close to a multilayer systems approach, trying to match organisational structures with decision layers in problem-solving; the book refers, in particular, to Simon.[8]

It is evident, however, that *participative planning and decision making* for large social systems, one of the emerging great themes of our time, finds an appropriate model only in a multilevel system, in particular in the multiechelon system. It is increasingly realised that long-range forecasting and planning for innovative social organisations require a basis of *decentralised initiative and centralised synthesis*[9]– the very principle underlying the multiechelon systems approach, although the subtle processes of understanding and self-motivation which are so essential for organising decentralised cybernetic planning, cannot yet be appropriately formalised. The notion of co-ordination and the theory of multiechelon systems thus become the *key to a new organisation theory geared to future-oriented systems of human decision making*.

Laying the foundations for a new theory, more appropriate for complex social and biological forms of organisation, is not the only important contribution of the book. It may be expected to trigger a real intellectual breakthrough in making clear *the distinction between basic stratified, multilayer (problem solving), and multiechelon thinking in general*. If multilayer decision thinking, so successful in its application to purely technological

---

[5] John von Neumann and Oskar Morgenstern, *Theory of Games and Economic Behavior*, revised edition, Princeton Univ. Press, Princeton 1955.

[6] J. Marschak and R. Radner, *The Economic Theory of Teams*, Cowles Foundation Monograph, Yale Univ. Press, New Haven 1969.

[7] Jay W. Forrester, *Industrial Dynamics*, MIT Press, Cambridge, Massachusetts 1961; *Urban Dynamics*, MIT Press, Cambridge, Massachusetts 1969; and *World Dynamics*, Wright-Allen Press, Cambridge, Massachusetts 1971.

[8] J. G. March and H. A. Simon, *Organizations*, Wiley, New York 1958.

[9] See Chapter 10 of this volume.

tasks, is turning into a gigantic failure when applied to the 'problématique' of social systems, which usually cannot be divided into clear-cut problems amenable to sequential modes of problem solving, stratified thinking characterises current political and economic approaches, and perhaps more generally any one of the single-dimensional approaches which fail to grasp the complex and systemic reality constituting the human world. Politically-oriented strata of the world system, such as 'world', 'power blocs', 'nation-states', 'autonomous regions or provinces', and 'communities', remain as isolated from each other in their descriptive concepts, terms, and overall systems views, as do economically-oriented strata of the same world system, such as 'world economy', 'economic blocs', 'national economies', 'corporations', etc. Stratified thinking is at the roots of the many *'dialogues des sourds'* taking place everywhere in the world.

Horizontally defined strata cannot be co-ordinated toward a common system purpose. Each stratum finds a purpose for itself. The purpose of a nation-state is not to improve, in any way, the world, but only the potentials for the particular nation-state itself. Power blocs and nation-states might, ideally, be co-ordinated toward a common world purpose. But, where some form of co-operation seems to become noticeable, it usually follows merely from some desire to form coalitions (or teams) to protect the 'egoistic' purposes of the nation-state. The result is single-level interaction lacking any view of an overall system purpose.

Terminology, principles, concepts and interfaces defined at a specific stratum, do not change when the stratum is recognised as part of a multilevel system; on the contrary, the whole system is interpreted in terms of specific stratum. As 'peace' is a notion for the 'world' stratum, so 'deterrence' is its equivalent at the stratum of 'power blocs', and 'political power, military strength, economic growth' are notions for the 'nation-state' stratum. Economic and technical aids are part of 'foreign policy', a notion for the nation-state; they are rarely considered at the 'world' stratum. The drastic failure of the First Development Decade of the United Nations is the result of stratified thinking – the rich countries defining 'development' in their own terms, and for their own purposes – as well as of multilayer thinking which assumes that there is a clear-cut economic problem which can be solved before other problems are tackled. And in the light of Forrester's recent coarse-grain investigation of 'world dynamics'[7] it even becomes altogether doubtful that the strategy of development through industrialisation can be justified at the 'world' stratum.

It is obvious that a pluralistic organisation of mankind, or of a particular society – embracing, for example, different cultural and ethnic groups – cannot be grasped by stratified thinking which allows only for some degree of freedom in horizontal organisation. The result is 'minority' problems everywhere, in communities, nations, and even at world level where non-Western cultures are becoming 'minorities' in spite of being more consonant (e.g. oriented toward cyclical development instead of growth, contemplation instead of competition) with a stable world system. Single-level thinking is conducive to single-goal thinking. . . .

Where a *purpose* of a whole system comes into focus – and a purpose is always pluri-dimensional – the question of co-ordinability of systems levels becomes decisive. In science, stratified thinking has not only served to cement the claim for 'purposeless' and 'value-free' science, but has also led to grotesque excesses of reductionist and behaviouralist thinking. If knowledge is understood, with Churchman,[2] as a way of doing or 'a certain kind of management of affairs', the orientation of science and other forms of human thought toward a purpose may be expressed through organisation in a multiechelon system.[10] Inter- and trans-disciplinarity may then be viewed as two-level and multilevel co-ordination of hier-archically ordered forms of thought. The paramount task of organising elements of technology, natural ecology, and the psycho-social sciences for integrative systems design in the human world, thus gains an entirely new perspective.

To understand the co-ordinability of human aims and ambitions at all levels, and for all purposes, emerges as the most important task of our time. *Local non-co-ordinability* among multiechelon system elements, be it within one dimension or in the interaction between different dimensions (technological/social, or political/economic, etc.), gives rise to *'problems'*. Overall *system non-co-ordinability*, for example due to the non-co-ordinability of different cultures at a high echelon of the world system, gives rise to a complex, inter-meshed *'problématique'* (as Ozbekhan calls it). This is the state of the world, and of mankind, today – and of many subsystems of human living. In this situation, the new approaches opening up with the systems theory conceived by Mesarović and his associates, constitute *an element of hope*. It ought to be taken up quickly.

[10] See Chapter 15 of this volume.

THE EMERGING 'WHOLE WORLD' PERSPECTIVE

12. A FRAMEWORK FOR INTERNATIONAL CO-OPERATION IN SCIENCE
    AND TECHNOLOGY
13. ROLES FOR THE 'WORLD CORPORATION'
14. GROWTH AND WORLD DYNAMICS

Science and technology have become dominant factors in social change, affecting – for the better or the worse – the whole world. Increasingly, problems have to be viewed at a global level rather than through the prism of regional, national, or corporate policies and interests. Technological innovation ought to be guided and stimulated with this perspective in mind.

Chapter 12 proposes a scheme for international co-operation in science and technology, following social rather than merely economic criteria. New corporate roles, viewed in a world-wide context, are developed in Chapter 13, emphasising social systems design as an emerging task for the 'world corporation'. Chapter 14, finally, in the framework of a book review on Jay Forrester's *World Dynamics* studies, deals with the implications of cancerous growth in a finite world, and the imperative to search for a viable transition from growth to balance.

Chapter 12 is published here for the first time.

Chapter 13 was first published under the title *'The "World Corporation": The Total Commitment'* in *The Columbia Journal of World Business*, Vol. VI, No. 3 (May/June 1971), and is reprinted here by permission of the Editors of the journal and Columbia University, New York.

Chapter 13 was first published in form of a book review on Jay W. Forrester, *World Dynamics*, in *Futures*, Vol. 3, No. 2 (June 1971), and is reprinted here by permission of the Editors of *Futures* and IPC Science and Technology Press Ltd., London.

12. A Framework for International Co-operation in Science and Technology

INTRODUCTION

IN the past two or three decades in which science and technology have become recognised as predominant factors not only in economic, but also in social development, our horizon has widened considerably both for the incentives offered by scientific and technological advances, and for the organisational principles best suited to plan and stimulate them in a rational way. Today, we see technology in the social domain act powerfully in two directions: substitution, and social change. Whereas substitution may be regarded primarily as economic process which is, to some extent, self-enacting and self-regulating, it is becoming increasingly evident that social change through technology also needs external stimulus and guidance to make use of the tremendous opportunities offered, as well as to avoid the pitfalls and dangers of a technology-dominated world. Such guidance seems the more urgent, since the one-sided pursuit of economic or other single-dimensional incentives, corresponding to narrow angles of view, has already given rise to serious distortions of the social framework.

Both the need for and the possibility of rationalising action through planning are accentuated by the ascent of technology as a dominant factor in social change. If technology is at the roots of processes such as world unification, population explosion, environmental deterioration, and the widening of various 'gaps' (especially the one between the rich and the poor), it is also capable of contributing to attempts to cope with these sets of problems if developed and applied at a higher degree of rationalisation than generally accepted today. Above all, it must be guided by a truly international, or even global, policy for science and technology.

International co-operative research and development has so far been conceived mainly in an economic framework. It is suggested here that this

narrow framework be abandoned for a wider one, namely social development at large, or even a global approach to the development of the roles
of mankind on Earth. International co-operation may find new aims in
applying spur and guidance to social and global development by means of
economic as well as technological, but also cultural and political, development. The partnership for such co-operation would be rooted in a common motivation and acceptation of the challenge to 'shape the future' of
mankind.

Accordingly, the criteria for selecting suitable programmes should not
be derived simply from economic incentives (such as task-sharing and
economies of scale, as important as they may be), but, in first line, from
the unavoidable necessity to co-operate on an international or even global
scale in problem areas affecting the whole world. Programmes will therefore have to be selected primarily in such areas where the need for global
policies and strategies imposes itself. Concrete examples are the World
Food Problem, population control, environmental technology, ocean
development, information technology, education, etc. Other criteria may
be the effect of the programme on social change and the chances to 'align'
action outside the co-operative effort to a considerable extent.

One may thus summarise the essential purpose of the proposed scheme
as that of performing and stimulating planning at a high level, and of
triggering action by taking the first step in the direction emerging from
such planning. It attempts to follow the succinct advice given by Lord
Rutherford: 'It is the first step that counts.'

THREE TYPES OF RESEARCH AND DEVELOPMENT

For the purposes of the proposed co-operative scheme, three types of
research and development, each one calling for a different approach at the
international level, may be distinguished:

The first one is *discipline-oriented fundamental research* of a type
necessitating large investment in research tools. In this area, fixed centres
of an international character still offer unique opportunities in various
respects: access for smaller countries to undertake research in areas from
which they would be otherwise excluded, rationalisation of complex
research programmes, and possibly considerable stimulation of scientific
entrepreneurship.

It is evident that such fixed research centres should be conceived only
in areas which are: (a) of 'permanent' interest, i.e. not only 'pregnant' with
potential significant results, but also underlying important potential
developments, especially of a kind that will lead to technological innovation contributing to social functions; (b) clearly definable in disciplinary
terms, possibly including inter-disciplinary combinations; and (c) suffi-

ciently wide to permit a flexible approach to research programmes and, possibly, inter-disciplinary interactions.

Suitable areas will probably include the following ones which deal with different aspects of the structure of matter: high-energy physics (where CERN stands already as the classic example), molecular (and sub-molecular) biology, certain sectors of solid-state physics and chemistry (especially where investigation under extreme conditions necessitates high investment in research tools – high neutron flux, high magnetic field strength, etc.), plasma physics, radio astronomy, possibly also brain neurology or 'biological information technology' in general (which may involve expensive experimentation with models, etc.).

It would be a matter of discussion to find out whether some of the inter-disciplinary new fields taking shape now - e.g. bionics – could gain much from being investigated in a fixed-centre framework. It is conceivable that the carefully aimed creation of 'melting pots' for disciplines may trigger and accelerate developments of a highly inter-disciplinary nature.

Apart from selecting the topics as well as creating and operating the fixed installation, an international co-operative scheme might have the continuous task to promote, guide, and co-ordinate the setting up of national 'satellite' institutes (a good example is provided by the Swiss high-beam-intensity accelerator, which has been conceived for roles complementary to those of the CERN extremely-high-energy accelerators). Although the fixed international instrumentalities will themselves become the natural centres of such a co-ordinated proliferation, it cannot be overlooked that the impact on national science policies will also reflect in higher-order co-ordinating tasks.

The second type is *flexible 'open-ended' research and development*, which notion is supposed to cover here research and development on a fundamental as well as on fairly applied levels, which is not in itself geared to specific technological innovations, although it will generally be conceived as contributing in a general way to potential innovation.[1] Typical results of this type of research will be elements of new basic technologies, or new knowledge accruing from exploration, etc.

This type of activity generally asks for at least a multi-disciplinary approach and is frequently characterised by quickly adaptable mission-

---

[1] The classical terminology, which would offer, for this category, notions such as 'oriented fundamental research', etc., does not fit the three-level scheme outlined here to distinguish between different requirements for the institutional approach. All research and development considered here is 'oriented' in one way or the other, and seen in a framework supposed to induce innovation. An example may elucidate this: While the structure of bio-molecules and the basic mechanisms of biological information transfer are investigated under the first type, the specific action of drugs in general, and the elements of a rational pharmacology, would be studied under the second type; the technological innovation of new specific drugs would then belong to the third type discussed in this section.

oriented patterns superimposed over a rigid discipline-oriented basis, a concept best implemented through a flexible task-force approach.[2]

It is certainly not desirable to create fixed instrumentalities in the form of large multi-disciplinary research centres which would comprise the resources needed for such an approach. Neither would it be possible to follow the diversity of problems required to give the centres a good 'load factor' (comparable to that of the research laboratory of a big industrial company), nor would it be feasible to achieve a desirable degree of flexibility.

It is clear that an international co-operative scheme cannot centrally initiate and control the flexible changes necessary for effective co-operation in this area. The organisational resources best suited to cope with this type of research and development are existing multi- or inter-disciplinary, and mainly non-industrial research institutes (e.g. of the not-for-profit type), provided that they are not too highly specialised. Concerning academic research, it may be preferable to deal with universities as a unit, not with university institutes. (American universities frequently have an internal research management office, a sort of 'brokerage' unit for contract research.)

The possible availability of laboratory chains, such as the American, British, and French national laboratories, would permit to assign the systematic exploration of multi-solution problems (such as new food production technologies), or the multi-track exploration of a given problem, to research 'super-units' which would themselves be in a position to act as 'control centres' to assure a flexible approach along the various lines of exploration as well as within the total research system for which they are given responsibility. It should be emphasised here that the real strength and entrepreneurship of such research 'super-units' has been stimulated chiefly when they were entrusted sizeable complexes of problems, that were at the same time clear-cut and well defined (e.g. nuclear weapons). Problem complexes such as the world food problem (particularly long-range aspects of new food production technology) may be expected to 'polarise' energies in the same way although elements of new basic technologies (e.g. photosynthesis in non-living matter, non-photosynthetic biological food production, etc.) are aimed at rather than technological innovation as such.

Possibilities of implementation include a wide spectrum from mere information exchange and loose co-ordination (the approach taken so far by OECD) to centrally funded temporary projects and more or less continuous project series. The more a recognised social function becomes the 'guiding force' for such a scheme, the more incentive there is for central funding. At the same time, the more complex and sizeable research and development, contributing to such a function, becomes, the less will it be

[2] See Chapter 10 of this volume.

possible to maintain flexible and adaptive control through central management. The schemes that will emerge, may therefore be expected to strike a delicate balance between the following extremes:

Decentralised initiative vs. Centralised selection
Large number of participants vs. Research 'super-units'
Rapid succession of temporary programmes vs. Continuous programme
Decentralised control vs. Centralised control

Suitable areas for international co-operation in this type of activity may include: Long-range aspects of food production technology (especially non-agricultural food production, space exploration (ESRO stands for a fixed-instrumentality approach, which is not encouraged here), ocean exploration (with emphasis on areas from which new technologies will emerge comprised under the notion of 'oceanics'), environmental sciences, certain sectors of medical research (e.g. a systematic exploration of cancer), certain sectors of the social and behavioural sciences, etc.

At the same time, certain fairly fundamental research areas might best be pursued by such a flexible approach, e.g. fluidics and certain fundamental aspects of the material sciences (such as a solid state theory of ceramics), and, in short, many areas of fundamental science and technology that may be raised to higher levels of understanding through a systematic and co-ordinated approach.

The third type, finally, in this enumeration, is *research and development geared to specific technological innovation*, with programme objectives either aiming directly at innovation (both hardware and software), or stopping short at the levels of basic technology or prototype development. Flexibility is of the highest importance here both for the selection of technological objectives and the carrying out of actual research and development.

In line with organisational concepts applied today to corporate planning in advanced-thinking industrial corporations, guidance to a permanent co-operative scheme may be envisaged to comprise: (a) long-range forecasting (in a 15- to 20-years time-frame, or beyond) and planning in a function-oriented framework with the purpose of preparing and evaluating the alternative strategic options (the technological decision-agenda); (b) the evaluation of alternative operational plans to attain specific technological goals (mainly within a 5 to 10 year time-frame); and (c) the final selection of technological objectives and the ways to attain them.

The notion of international co-operation should be geared primarily to international utilisation of the results, not at international set-ups for the actual programmes. In many cases, it will be preferable to give one research institute, or one corporation, responsibility as prime contractor for a clear-cut programme and for achieving a clear-cut technological objective. Where consortia do not form in a 'natural' way, they should not be forced.

This third type of research and development activity may be expected not only to become the biggest and most prominent part of a fully fledged international co-operative scheme, but also to stimulate and 'align' large areas of industrial activy in general.

As a matter of principle, mixed public/private funding should be envisaged as far as possible for programmes leading to innovations in marketable technology. If the co-operative scheme incorporates an effective guidance function which radiates far beyond its institutional boundaries, it may even be considered to establish an 'open' scheme which would attempt to deal flexibly with a wide spectrum of public and mixed public/private funding – extending to conditional forms of public engagement such as repayable credits or guarantees.

INDUSTRY AND PUBLIC INTEREST

Until recently, there has been relatively little interaction in Western countries between public interest and industry outside the defence and space areas or other fields in which the government represented the sole or principal customer. With 'big science' as a basis for industrial activity, exemplified by the development of nuclear energy, a public responsibility for partial funding has become recognised. Frequently, the main incentive was the strengthening of national industry in a specific technological area. Other national co-operative schemes took either the form of encouragement of the formation of national industrial consortia with public support (e.g. the British NRDC scheme), or of the formation of mixed public/ private companies (Italy).

International co-operation, with public support, is exemplified by the British/French 'Concorde' supersonic aircraft, and by some Euratom and OECD projects. Criteria so far were derived mainly from national or supranational motives, or considerations of market size, etc. However, hardly has a *creative* contribution ever been asked for from industry. Rarely has industry so far participated in planning at higher than the operational, or occasionally the strategic level.

The interaction between the consumption and the process activities – the principal 'axis' of social creativity – is becoming more intense today, as the process activities move closer and closer to the consumption activities (growth of the tertiary service, and more recently a quarternary 'information' sector). The integration of demand and response has been pushed farthest by industry, mainly by employing information technology as the link to the customer. On the one hand, this has given rise to a short-circuit between demand and response – with industry itself generating artificial demands (the 'waste-makers', the military–industrial complexes, etc.). On the other hand, industry is increasingly becoming aware of genuine social motivation and is trying to tie its own corporate objectives in with national and general societal objectives. Already, industry is becoming active in areas such as 'education', 'hospital management' (or health services and systems in general), 'city building', and 'regional development'. This trend is also implied in the establishment of long-range strategic forecasting and planning in a function-oriented framework.[3] It may be expected that, in advanced countries, industry will emerge as a major 'planner for society' in the 1970s.[4]

Industry, geared to a rational and well-focussed innovation process, is by far best suited to cope with the whole range of problems encountered in technological innovation also in such areas which are characterised by public interest. This implies that industry should not only be utilised to develop and supply technological expertise and build components to specification, but should also be given responsibility in the entire spectrum of contributing activities, especially the following ones:

—Software studies, in particular systems analysis studies which attempt to assess and evaluate the strategic technological options for broad social functions, and the tactical options for solving specific problems.
—Prime-contractorship for the development of complex technological or socio-technological systems or basic technologies.
—Participation in the above as subcontractor.
—Contractual research of a more 'open-ended' type, with emphasis on specific types of investigation rather than on technological objectives (although this should perhaps be the least important role for industrial participation).

Strong and effective incentives for industry can be provided only by short-range and long-range (strategic) economic advantages (*see Table 12.1*). It is clear that public funding of development projects may distort the 'free enterprise' system, especially where the public is not the only customer. Such a distortion has to be limited to the extent that the primary creative factor, namely industrial entrepreneurship, is to be maintained. To this end the following areas and rules for public funding of industrial participation in an international co-operative scheme may be adopted for flexible application:

—Contributions to problem complexes which are too big to be tackled by individual companies, and which can be solved only in line with a large-systems approach should be favoured.

[3] See Chapter 8 of this volume.
[4] See Chapters 9, 10 and 13 of this volume.

Primary objective of co-operation	Incentive for co-operating industry	Preferable way of funding
Systems analysis	Conceptual lead	Public or mixed public/private
Basic technology (hardware or software)	Option to enter systems development in favourable position; 'entrance fee' partly paid for; training of people	Public or mixed public/private
Prototype of technological system	Lead in marketable technology	Mixed public/private
Supply of technological system or service		
(a) Public sector is principal potential customer*	Temporary monopoly	Public or mixed public/private
(b) Private sector is principal potential customer	Market dominance and strategic position	Private with some public encouragement (guarantees, credits, etc.)

* Only for this objective of co-operation a relationship similar to that in defence and space research may develop.

Table 12.1. Incentives for industry, and modes of funding.

—New fields for industrial activity, where no well-established competitive structure exists, should be emphasised.

—Developments for which the public sector is the principal potential customer should be favoured; this will be the case primarily in the connective and organising activities (transportation, communication, specific developments in information technology, etc.)

—The share of public funding ought to be more substantial for systems analysis studies and the development of new basic technology, than for the development of marketable technology.

—Competition is completely open for the various steps of a given development; e.g. if company A (or, preferably, a number of companies) is selected for the systems analysis step, company B may be awarded prototype development, and company C prime-contractorship for the supply of the resulting technological system. At each stage the solutions offered for the next step are evaluated anew.

—Efficient project management is considered an asset.

—Incentive-type contracts are used as far as possible.

Such ideas as 'strengthening industry' in certain fields and/or certain regions are a matter of national or supranational policies and should not enter the considerations developed here. Of course, national policies may be combined to pursue such objectives on a bilateral or multilateral basis, but their motivation is outside the notion of international co-operation adopted here.

It is essential that a dynamic view of the relationship between industry and the public sector is adopted in all areas where co-operative schemes are started. Such a dynamic view should already be incorporated in the systems analysis stage of a new development. The danger of neglecting the long-range view, including the situation after the objectives have been achieved, is demonstrated by the examples of several national nuclear energy programmes that lost their principal objective when the goal of economic nuclear power was achieved outside these programmes. It is of the utmost importance that the relationship between industry and the public sponsors before and after achieving the objectives is clear from the beginning.

In areas of considerable social impact, which provide the most 'natural' opportunities for international co-operation, one may think in terms of three distinct phases:

(1) Public motivation and funding guide industrial development activities toward new problem areas, and promote the search for new technological solutions.
(2) The new solutions are introduced into the social system by means of public organising activities.
(3) The new technology becomes an industrial growth and profit area.

Public action in the first two phases will be the more necessary, the more social change and the transformation of large systems is aimed at (world food problem, education, communication, etc.).

It should be noted that many of the technological developments envisaged here imply the generation of new industrial functions and new inter-disciplinary branches of industry. There is a strong mutual interest that these functions are, on the one hand, actually developed, and that they are, on the other hand, accepted in the social system. There is, for most issues at stake, no alternative to an *integrated* approach of public organising and industrial process activities. With the alignment of industry to social functions, the former strict separation between private enterprise and government cannot be maintained.

The future pattern of international co-operation will probably see industry participating in many different roles. It can be expected that the intiative taken by the public in certain areas, and the guidance provided by systems analysis studies and corresponding public engagements, will stimulate industry to offer contributions on a fully competitive basis, not supported by public funds. International co-operation may thus promote an 'alignment' effect which is most desirable. It will be an important task for the international co-operative scheme to allow for such 'competition' from the outside – even if a publicly funded project should be completely out-manoeuvred by private initiative – and to offer it equal opportunities for introduction into the societal systems.

THE ORGANISATIONAL ELEMENTS OF AN INTERNATIONAL CO-OPERATIVE SCHEME

The following overall criteria may be adopted for the establishment of an organisational set-up to cope with the problems of international co-operation in science and technology:

—It is organised so as to permit planning and programming in a function-oriented framework, and is flexible in the engagement of scientific and technological disciplines and approaches.

—It makes use, as far as possible, of decentralised initiative in the planning and programming stages as well as in carrying out actual research and development programmes.

—It is effective, restricting public (centralised) initiative chiefly to the linkage of technological tasks to social functions, and public control essentially to budgetary control.

—It permits a wide spectrum of approaches to a mixed public/private partnership, which, as a general rule, should form the basis of the co-operative scheme.

—Action supported by the co-operative scheme should, as far as possible, stimulate an avalanche of private or nationally supported activity outside the scheme, taking the same direction within an overall systems approach.

—The spreading of special skills, developed in connection with work supported by the co-operative scheme, is stimulated, for example through mobility of people.

—Implementation of the results, where feasible (e.g. the introduction of a specific new technology) is promoted on as broad a basis as possible by assuring access to the results; where the public sector is responsible for the implementation, suitable links to governmental or other bodies should be established already in the planning stage.

A simple institutional framework (*see Figure 12.1*) may be envisaged for an initiative stage in which neither the volume of programmes nor the number of complex systems to be dealt with are yet very large.

Apart from existing organisations and one or the other newly created fixed research centre, this simple scheme incorporates only two new elements: a systems analysis and guidance function, and function-oriented administrative centres.

The envisaged international co-operative scheme should be implemented by following rational decision making starting at policy level. The overall objective of the scheme – to apply spur and guidance to the technological aspects of social development – and the ambitious selection

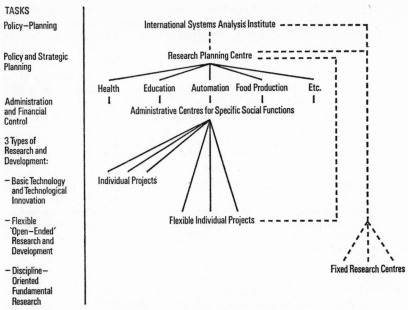

TASKS

Policy–Planning

Policy and Strategic
Planning

Administration
and Financial
Control

3 Types of
Research and
Development:

– Basic Technology
and Technological
Innovation

– Flexible
'Open–Ended'
Research and
Development

– Discipline–
Oriented
Fundamental
Research

Fig 12.1   A schematic organisation framework for international
co-operation in research and development

criteria, call for a sophisticated *systems analysis and guidance function* at
two levels:

(a) A 'Research Planning Centre' would take a comprehensive and long-
range look at social development and the implications for technological
development in general. Such a centre could best be incorporated in an
international systems analysis institute (a 'look-out' institution, or an
'Institute for the Future').[5] The overall task of such a 'look-out' institution would be to explore, assess and evaluate alternative 'possible futures'
and consistent sets of technological elements, and to map out alternative
paths to achieve them. It would, in Dennis Gabor's words, 'invent the
future'. Against this general background, the incorporated 'Research
Planning Centre' would attempt to formulate policies and evaluate
high-level missions for technological development, develop criteria for
the assessment of proposed solutions, determine priorities, and specify
areas particularly suited for promotion through international co-
operation.

Ideally, the 'Research Planning Centre' would attempt to provide
guidance for the overall technological development contributing to
certain social functions, not just to the sectors selected for international
co-operation. Full success of the idea of international co-operation can
only be achieved when national and private initiative is stimulated to a
large extent along the same lines which are to be promoted through

[5] An International Institute of Applied Systems Analysis (IASA) is under active
consideration by a number of Western countries and the Soviet Union.

public funding and co-operative projects. Total research and development required to attain recognised social goals may well amount to between ten and a hundred times the amount of public funding that can be expected.

It is therefore necessary, that the 'Research Planning Centre' works as far as possible 'in the open', disseminates its findings publicly in an effective way, and tries to become an element of public thinking and public opinion. This accent on 'openness' will also be required for the stimulation of decentralised initiative at the second level of systems analysis, to be outlined below.

The 'Research Planning Centre' will finally be responsible for the evaluation of systems analysis studies provided on level (b), and for the overall conceptual co-ordination of projects.

(b) Decentralised initiative should be aimed at for systems analysis studies linking missions under social functions to alternative technological developments. In general, these studies should be treated in the same way as actual projects in respect of public participation in the funding and the availability of results. The multi-disciplinary not-for-profit research institutions, which have pioneered this type of software research, and advanced-thinking industry (exemplified by the American aerospace industry) both offer the best available capacity for this task. They are also in the best position to provide tentative planning for alternative paths to attain specific technological objectives or missions – a creative input which, in turn, can be expected to influence the assessment of mission priority and funding level, or even the overall balance between recognised social functions.

This two-level guiding function will deal mainly with medium- and long-range aspects of a function-oriented framework in areas of public interest. A main difficulty will arise from the fact that, apart from the United States, few countries have yet adopted elements of function-oriented thinking in their governmental budgeting system; only the French National Plan may be adaptable to some extent.

On the other side, as has been pointed out above, the proposed scheme for international co-operation should not be treated as a 'resource allocation problem', i.e. as part of the problem on a national level how a bulk sum for the support of scientific and technological activity should be allocated. It should, on the contrary, be regarded primarily as a contribution to the various problem areas of social development which governments face on a national level.

It might therefore be envisaged for a transient period that the 'Research Planning Centre' attempts to work out a partial function-oriented framework for the governments of the participating countries and proposes, on this basis, flexible funding schemes in form of a 5- to 10-year sliding plan (something like an 'inter-governmental Planning-Programming-Budgeting System' for some of the most important social functions). Although such a medium- to long-range proposal can obviously never be formally accepted, as long as governments adhere to their present short-range instrumental budgeting schemes, the adoption of such non-obligatory guidelines could become most valuable as an indicative basis. If short-range budgetary measures by the individual governments ruin this plan, it would at least be clear where and how social development will suffer from it.

Providing, for the co-operative research and development scheme, a function-oriented framework for the benefit of government planning and budgeting, may also, hopefully, act inducive to governments to provide funds for the scheme outside their national research budgets, and thus to avoid an either/or attitude which would only have the effect of creating strong opposition for the international co-operative scheme from the side of the traditional receivers of national research funds.

The other new organisational element, the *Administrative Centres*, should preferably be set up separately for each social function dealt with. This more or less stable function-oriented structure will then represent a compromise between a central 'Research and Development Authority', or 'Science and Technology Agency' – an instrumental approach which, as we learn today, presents almost insurmountable obstacles to long-range strategic and functional thinking – and a flexible pattern of technological mission-oriented agencies, which it may be impossible to establish at the international level where changes come slowly and involve complicated negotiations.

The Administrative Centres are supposed to develop initiative separately and 'in competition' with each other, and to try to win public and private support for their 'businesses' (their programmes). They will do so by offering contributions to their respective social functions in the light of the overall futures-planning of the International Systems Analysis Institute.

A typical 'flow-scheme' of the process of planning, decision making and action, indicating major responsibilities, may then be envisaged as shown in *Figure 12.2.*

Each contribution to this process would be negotiated separately, so that, for example, different research institutes or industrial companies account for the systems analysis, research aiming at new basic technology, and research and technology geared to technological innovation, all within the same functional programme.

Certain top-level co-ordinating tasks, which are still outside this scheme, could well be taken care of by the usual Ministerial Meetings held at OECD, or parliamentary discussions in the Council of Europe: Approval of recommendations formulated by the International Systems Analysis Institute, creation of new Administrative Centres and adoption of new functions to become part of the co-operative scheme, the general level of support, ultimate financial control, etc.

It may become necessary to establish *ad-hoc* 'utilisation programmes'

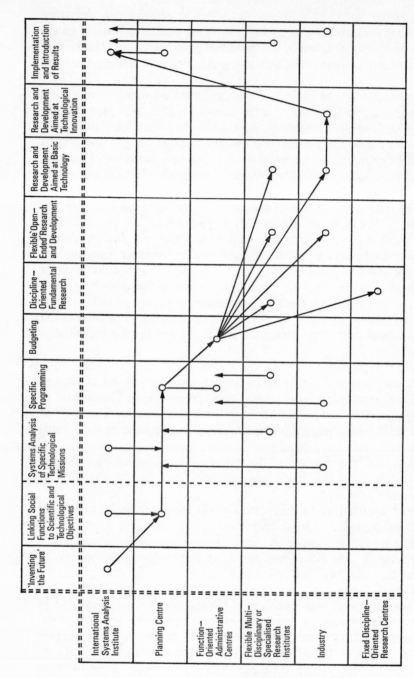

Fig 12.2 Typical shared responsibilities in the co-operative planning, decision making and action process.

where co-operative activities result in basic technology or potential technological innovation. This will be the case particularly where the introduction of a new technology in the social area is:

—part of a complex (international or global) system of social change;
—subject to government or other national-level action;
—connected with high costs;
—opposed to the direction in which the economic and social systems move by their own inertia.

Wherever possible, however, the free use of the results should be encouraged by treating them as completely 'open information'.

Basically, research and development performed under the auspices of the co-operative scheme, is supposed to create a strategic impact, set the direction for scientific and technological activity *outside* the co-operative scheme, and 'trigger' an avalanche of activity in this direction.

13. Roles for the 'World Corporation'

RESPONSIBILITIES, NOT RIGHTS

INTERNATIONAL business, so far developed mainly in a variety of modes corresponding to the exploitation of specific rights granted by a liberal economic-political system. These rights, in particular, permitted the extension of relations between producer and consumer beyond the boundaries of nation-states, in other words, the superposition of an international economic system over a multitude of national or regional systems. This creation of a supersystem which is made up of parts of national economic systems, did not follow any global design or general anticipation – not even to the extent to which the structures of free market economies are anticipated and designed in many liberal countries.

The result is something new: A system essentially lacking any explicit mandate and mechanism for searching and defining an overall policy, i.e. norms and values which are in consonance with its own *problématique* (as Ozbekhan[1] defines 'policy'). The fragile mechanisms of international monetary, trade and tariff control and the various consultative bodies constitute, to the extent that they are effective at all, merely some weak negative feedback control applied to selected economic parameters. This is enormously insufficient for the role international business plays in determining the future of the world in respect of exploitation of resources, economic and social development, and inter-cultural relations.

Most of the rules of behaviour for international business are set in a fragmented way by nation-states, reflecting not the concern for the overall international system or a world system, but the narrow interests of subsystems. Adapting to these varying rules, which rarely go further than setting a 'management by exception' pattern, the multinational corporation has developed typical responses which are summarised by Perlmuter's[2] categories of ethnocentric, polycentric, and geocentric management. These categories span a wide spectrum from a high degree of

[1] Hasan Ozbekhan, 'Toward a General Theory of Planning', in E. Jantsch (ed.), *Perspectives of Planning*, OECD, Paris 1969.

[2] Howard Perlmuter, 'The Tortuous Evolution of the Multinational Corporation', *Columbia Journal of World Business*, Vol. IV, No. 1 (January–February 1969).

centralisation in the home country, over a delegation of lower-grade responsibilities to local management in many countries, all the way to the concept of top management by a cosmopolitan group of co-ordinators of activities in many countries. Only Perlmuter's geocentric type lets the idea of a 'world corporation' become vaguely visible. However, it is this idea which ought to move into the focus of attention today, because it promises a possibility to design policies which take the whole world into account – a possibility which is generally not realistic for the existing political structures, including the United Nations and its agencies which are geared to the classical process of political bargaining rather than to searching and forming viable policies for the world. International business has been the most successful institution in transgressing the boundaries of nation-states and even of power blocs. It is now called upon to become a leader in the development of world institutions, which will eventually also include effective political institutions – but the role of business will be crucial in the transitory period.

What is a role? Is it just the exploitation of rights, as business traditionally seems to assume? Geoffrey Vickers, in his recent book,[3] remarks that 'rights do not make a role; they are at most ancillary to the status which it carries. Responsibilities, not rights, give meaning to personal and political life. . . . A declaration of responsibilities implies far more clearly the conditions needed to attain the rights. . . . Though they do not state a programme of action, they define a resetting of actual regulators.' The definition of roles may be viewed as the main organising principle leading from the formation of policies to the creation, or renewal, of institutions.[4] The cybernetic cycle between values, norms, policy formation (which amounts to writing a 'declaration of responsibilities'), the definition of roles and the creation or renewal of institutions may be viewed as sketched in *Figure 2.3*.

Thus, business ought to define, and continuously redefine, its roles and its institutional structures in the light of 'total system dynamics', i.e. through an understanding in which direction our dynamic situation will develop if specific norms and regulators are brought into play. The institution of business, here, is itself one of the regulating mechanisms. Policy-planning, in a business context, means nothing else but reflecting on the corporation's role in a changing environment – where this environment is the entire milieu with which the corporation interacts, the social, political, economic, technological, psychological milieu, not just the market place. Such a 'top guiding task' for corporate planning is about to find its place in the structure of corporations operating in fields of considerable social impact.[5]

[3] Geoffrey Vickers, *Freedom in a Rocking Boat: Changing Values in an Unstable Society*, Allen Lane, The Penguin Press, London 1970.
[4] See Chapter 2 of this volume, in particular, *Figure 2.3*.
[5] See Chapter 10 of this volume.

THE CHALLENGE OF REGULATION

Before exploring how the 'world corporation' might find its proper role, it is perhaps illuminating to first try to clarify the nature of the 'total system dynamics' which is to be understood, and then to look what form the response of business is taking in a national, or local market, context – to the extent that such a response is already becoming visible.

Vickers, to whom we owe perhaps the clearest explanation of the cybernetic process of history and the consequences for humanly dealing with a systemic world,[3] sees regulation breaking down today in four main fields: *ecology* – the relations between ourselves and the physical milieu; *economy* – the relations between our activities as producers and consumers; *politics*, in the widest sense – the relations between ourselves as doers and as 'done-by'; and *appreciation* – the inner coherence of our systems of values and norms (where the notion of values comprehends at least interests, expectations, and standards of judgment). This is a very useful frame of reference which will help to avoid discussing regulation merely in economic terms, or at best in economic/ecological terms which is as far as conventional wisdom is prepared to go today.

Business, and above all business dealing with advanced technology, has to do with all four areas in which regulation is breaking down. In highly industrialised countries, regulation in ecology and politics is becoming of greater concern today than regulation in economy, and the gap between young people's values and the 'establishment's' is bringing the issue of appreciation to the foreground. In a world context, the breakdown of regulation in economy (the growing gap between the rich and the poor countries) and appreciation (the imposition of Western values, technology and development patterns on other cultures and civilisations) constitutes the most burning issue, with ecology and politics coming second only in the fight for subsistence in which the larger part of mankind is involved.

Business operates in this situation which is characterised by a desperate need for regulation – a regulation which still has a chance of being established in a decentralised, cybernetic and free way, not as rigidly imposed control from the top. Business, as one of the potentially most creative institutions of society – and, together with government and the university, the 'manager' of science and technology – finds itself at a crossroads today. It can take an active or a passive attitude *vis-à-vis* the task of 'planning for society', of participating in the regulation of the world system and developing a suitable role within such a policy. It can go on interacting with fragmented social systems, such as markets, and act itself as a fragmented social system pursuing the goals inherent in its own structure, or it can become a leader in a movement toward a more fully integrated society.

The decision to be made now will in the long run determine the life or death of the business corporation as a free agent in the psycho-social evolution of mankind.

Current types of responses at national level include attempts to establish 'technology assessment',[6] looking at foreseeable impacts of potential technological options on the systems of human living – in other words, to make technological forecasting at strategic level a publicly stimulated and co-ordinated task. The weakness of this approach lies in the lack of mechanisms providing normative guidelines and criteria to be applied to matters of choice, such as alternative technologies. It can at best lead to something like a 'free market for technology' which is curbed by legal constraints established *ad hoc* wherever bad 'side-effects' are recognised in anticipation or *post mortem*. Again, like for the operating framework for international business, some passive 'management by exception' is envisaged here and its norms are established in a fragmented way only. The complex relations between economy and ecology, it is proposed, are to be dealt with by a single-dimensional economy/diseconomy approach, or, as the head of an inter-governmental organisation put it recently, the subjection of technology to 'the superior thinking of the economists'.

Of far greater importance are conscious efforts in business organisations, in particular in industry, to recognise functions, or broad need categories, for technology in society more clearly and to organise business activities around such functions rather than around the exploitation of specific raw materials or skills, or around specific technologies or product lines.[5,7] Examples of such broad functions would be power generation and transmission, telecommunication, transportation, information, health, food production and distribution, education, etc. It can easily be seen that focussing on such functions implies gaining flexibility in respect of the means (technologies, products) to be applied. Calling this emerging trend in business 'problem-orientation', as is sometimes suggested, grasps only part of its essence. The task here is not to solve problems once and for-ever, and one after the other, as in dealing with complex technology, but to invent and design parts of viable socio-technological systems, to engage in 'social architecture' as Perlmuter[8] has called it. This means that business activity is no more only economic or economic/ecological in nature, but includes also the dimensions of politics and appreciation, in other words the totality of Vickers' scheme outlined above. Such an

[6] *Technology: Processes of Assessment and Choice*, Report of the National Academy of Sciences; and *A Study of Technology Assessment*, Report of the Committee on Public Engineering Policy, National Academy of Engineering; both reports published by the Committee on Science and Astronautics, U.S. House of Representatives, July 1969.

[7] See Chapter 8 of this volume.

[8] Howard Perlmuter, 'Corporate Markets and the Human Condition', *Innovation*, No. 9 (January 1970).

integrative approach, in particular in the development of technology,[9] will have a natural built-in tendency to link up corporate policy with societal policy, to 'reach up' to the level where the dynamics of societal systems become of concern, and thus to make business play a political role, in the broadest sense of the word, as a 'planner for society'. With this definition of a new role, or even a complex role structure, in a societal context, the business corporation participates in the cybernetic cycle and engages actively and consciously in shaping mankind's future. This stage has not yet been reached, but a start has been made in this general direction and awareness of the necessity of new roles for business is increasing and gradually changing business attitudes.[10]

At national level, there exists a mandate for regulation. It is entrusted to government, whether dictatorial or democratic. But it may already be noticed that the design of viable complex systems of which, in particular, society and technology are constituents, requires a degree of inventiveness and creativity which rules out any possibility that government can play any but a synthesising and co-ordinating role. The modern business corporation is moving away from a hierarchical organisation in which ideas and plans were handed down from the top, toward approaches characterised by *decentralised initiative and centralised synthesis* – something analogous has to happen if it comes to redesigning society's systems and marshalling the resources of science and technology to that end. The inter-linking of new roles for government, business, and the university may be expected to emerge.[11] These are developments which still lie in the future.

At international level, however, no mandate for regulation is given in any but the weakest and non-committal forms. The international corporation has already become the most effective agent in the transfer of technology – but also of Western economy, Western culture and civilisation and Western attitudes toward ecology. It will find opportunities to participate in the regulation of the world perhaps in a more natural way even than opportunities to develop a role within a nation-state. A 'world corporation' defining its role in this way would go beyond Perlmuter's geocentric corporation which is also the type held in mind by Ash.[12] It would also go beyond the global problem-solving corporation which Barber[13] advocates (e.g. a world weather forecasting corporation). Its vision would embrace more than just 'the world as a total market'.

[9] See Chapter 3 of this volume.

[10] *Perspectives on Business, Daedalus*, Winter 1969, especially the contributions by Neil W. Chamberlain, Michel Crozier, Eli Goldston, and Leonard S. Silk.

[11] See Chapter 9 of this volume.

[12] Roy L. Ash, 'The New Anatomy of World Business', *Columbia Journal of World Business*, Vol. V, No. 2 (March–April 1970).

[13] Arthur Barber, 'Global Problem Solving – A New Corporate Mission', *Innovation*, No. 15 (October 1970).

TWO APPROACHES TO WORLD INTEGRATION

Before going further, an essential distinction has to be made. Integration of world-wide activities may mean two very different approaches for the 'world corporation', and the decision between them will be of crucial importance:

—One approach consists of imposing Western patterns and Western dynamics of economic, industrial and implicitly also political, social, and cultural development on the rest of the world. It is increasingly recognised that such a policy is inherently disastrous – if the present world population could live at the average technological level enjoyed by the United States citizen, this would mean a seven-fold increase in the use of the Earth's resources – and it is already consciously rejected even by representatives of poor countries.[14] Nevertheless, it remains so far the only direction in which the international and the forerunners of a 'world corporation' have made significant headway.

—The other approach would attempt to co-ordinate development efforts geared to different cultural patterns in such a way that the world can be regulated as a 'trans-cultural system'. This approach, which corresponds to the real role of the 'world corporation' as viewed in this chapter, leads to the questioning of our Western cultural basis itself. It is just becoming barely visible in some of the attitudes developing in business, but the consequences are so penetrating that few people as yet dare to discuss them seriously and in an unbiased way.

Within the *'Western' approach*, the traditional 'colonialistic' attitude of restricting the innovation process to the home country and of building merely production places in a strategic pattern around the world – focussing on a market concept of the world – still holds strong. It may be identified with the 'low-integration' notion of a multinational corporation. Only where *innovation* is planned and carried out through international decentralisation – which may be identified with the attitude of a truly international corporation – a genuine (if tentative) role for the 'world corporation' has been recognised and developed, generally according to one of the following models:

(a) A network of corporate-level laboratories, spread over a number of countries, and directed and staffed primarily by local scientists and engineers which interact directly with corporate management (e.g. CIBA, Shell, Unilever).

[14] Report of the conference 'Technology: Social Goals and Cultural Options', sponsored by the International Association for Cultural Freedom and the Aspen Institute of Humanistic Studies, Aspen, Colorado, 29 August–3 September 1970, *Innovation*, No. 15 (October 1970); the same trend also became visible in the Second International Conference on the Problems of Modernisation in Asia and the Pacific, Honolulu, Hawaii, 9–15 August 1970.

(b) Interaction between a restricted number of corporate-level laboratories and a large number of laboratories owned by individual member-companies of the international group, sometimes reinforced by regional co-ordination offices. A typical local laboratory will carry out activities along three lines: adapting technology available in the 'corporate pool' to local requirements, improving 'corporate pool' technology by its own developments, usually triggered by specific local needs, and finally participating in a co-ordinated, or sometimes deliberately competitive, effort to enrich the 'corporate pool' in an important way (e.g. ITT, Philips).

(c) Interaction within a laboratory pattern similar to (b), but in such a way that the innovation process is co-ordinated centrally and managed separately from all other affairs through a network of communication channels tied together at corporate level (e.g. IBM).

From a 'Western' angle of view, such internationalised processes of technological innovation are already very effective in many ways: The corporation may tap creative resources (people, skills) not transferable to the home country; skills in advanced technologies are developed in countries lacking them, the brain-drain finds a remedy, and local activities in science and technology are usually stimulated far beyond the areas taken up by the corporation.

A *'transcultural' approach*, in contrast, would start by looking at the world system and the dynamic situation in which it is trapped, at the instabilities developing at global scale, and the breakdown of self-regulation.[15] It would then look at the impact of various cultural values, attitudes and standards on the world system, and above all at the distortions inflicted on it by the cancerous behavioural patterns of Western society. Again in the conceptual framework of Vickers' scheme, some of the conclusions reached may then read as follows:

In the *ecological* area, regulation is endangered by the Western attitudes to dominate Nature, exploit its resources wherever single-dimensional economic criteria are believed to warrant it, generate waste through open-cycle economies, upset natural cycles and introduce new types of feedback interaction between man and his environment to a potentially dangerous extent.

In the *economic* area, regulation is endangered by the growing gap between the rich and the poor countries, by Western growth patterns and stagnation elsewhere, by a growing mismatch between and separation of production, distribution and consumption activities, by 'technology gaps', and by the distortion of the reward system. Whereas raw materials remain stagnant in price, or become even cheaper as a result of competition with

[15] Aurelio Peccei, *The Chasm Ahead*, Macmillan, New York 1969; significantly, the author is one of Italy's best known industrial managers.

the artificial materials entering with technological advancement, price increases benefit primarily finished products, and the service sector. The almost total failure of the United Nations First Development Decade bears out this basic insufficiency of the reward system. The Western 'self-exciting' economic system (as Vickers justly calls it) cannot go on in the same direction much further. But, where appeals to self-restraint rarely go beyond asking for some attenuation in the rate of growth, what actually may become necessary is a voluntary lowering of Western material standard of life (and waste) if the task of establishing some viable world balance is taken seriously!

In the *political* area, regulation is endangered by the powerful linear development trends, increasing complexity and 'clogging' of the systems of human life, and threats to human freedom and dignity which are unleashed by the indiscriminate development of technological potentials. The cancerous nature of this development is also augmenting other cancers, such as population growth. If the systemic interactions of technology are becoming of major concern now for Western society which carries technological development, they are causing even greater distortions in those parts of the world whose cultural basis until recently engendered only modest uses of technology.

In the *appreciative* area, finally, the growing dominance of Western values not only endangers a cultural plurality in the world, but is at the very roots of the breakdown of world regulation. As Ferkiss[16] points out, Christian thought knows a beginning and an end of the world, and in between 'progress'. In life, we have to achieve something – and we measure it usually in terms of growth of standard of life, knowledge, and numbers. This also leads to overemphasis on the value of competition, from individual competition – leading to increasing alienation, neurosis, and even violence – over-competition between enterprises, to competition between nation-states up to various forms of war. Responsibility, which would lead to the formation of role structures capable of acting as regulators, does not yet assume due prominence. On the other hand, other cultures and religions (e.g. in Asia) know a completely different basis of thought and action, closer to balance and cyclical development. For them, achievement is not a dominant idea for life on Earth, and work not a value in itself. Where Christian man, expelled from Paradise and condemned to work, has made work his new Paradise, Asian values have favoured learning and contemplation as man's Paradise on Earth. The results, it must be said, have in some instances been degradation into stagnation, failure to free man from his most basic and vital needs (food, shelter, clothing, etc.), and a desperate fight for new, richer, environments as long as they were not in short supply. Inter-cultural exchange in the form of Western technology

[16] Victor C. Ferkiss, *Technological Man: The Myth and the Reality*, George Braziller, New York, and Heinemann, London, 1969.

upsetting the population balance in less dynamic civilisations of the world, has dramatically increased this calamity.

CO-ORDINATION INSTEAD OF DOMINANCE

The 'Western' approach has fully worked only in the case of Japan which has abandoned her own cultural basis in record time and adopted Western standards as articles of faith, making them even more incisive than in the West. To many observers, however, the outcome of this cultural 'heart transplant' is still uncertain, the price for economic success possibly very high. The point, however, is that joining the Western bandwagon is simply impossible for the whole world – because the Western subsystem has soared beyond any level tenable for world stability.

Cultural plurality is a most important asset in this transitory situation in which gross distortions of regulators and in the interaction between subsystems have to be remedied. The value of such a pluralistic approach is also becoming evident in the relations between the United States and Europe, and even between the United States and Canada.

The key to the maintenance of cultural plurality is *co-ordination*, which will have to replace dominance. Co-ordination is not to be confused with centralised rigid control, or enforced co-operation. It means bringing into play energies and ambitions at *all* system levels. Not national or Western or any other subsystem views should govern this co-ordination, but the common purposes of mankind – which, in many instances, are yet to be clearly defined. The corresponding systems notion is that of a *multiechelon* or *multilevel, multigoal system*,[17] which is 'alive' in all its elements. The direction of the overall, the world system, would then not be determined by any subsystem or co-ordinator, but by the aspirations of all parts at all levels of the system to the extent that they can be co-ordinated toward a common aim: cultures, their indigeneous institutions, specific instrumentalities, etc., down to individuals.

Currently, economic thinking is generally not co-ordinated in such a way, neither within a world, nor within a national system. Rather, it represents so-called *stratified* hierarchical thinking, with different sets of principles, axioms, concepts, and views of the overall system objective to be found at each of the strata: economic blocks, national economies, corporations, divisions, etc.

With these ideas in mind, *the primary role for the emerging 'world corporation' may now be seen in the co-ordination of a transcultural, i.e. total-mankind-oriented approach to shaping the systems of human living.* In

---

[17] Mihajlo D. Mesarović and D. Macko, 'Foundations for a Scientific Theory of Hierarchical Systems', in L. Whyte et al. (eds.), *Hierarchical Structure*, Elsevier, New York 1969; see also Chapter 11 of this volume.

planning and implementing ways of co-ordination, the 'world corporation' will act as a *world institution* – the first real one, if weak attempts to plan and co-ordinate by some of the United Nations agencies (e.g. FAO, the Food and Agricultural Organisation) are discounted here. This new world institution will need a wide variety of instrumental approaches to be carried out by individual, and to a certain degree competing, corporations which accept the basic role of the 'world corporation'.

First attempts to recognise such a co-ordinating role may already be found, for example, in regional development tasks (Greece, Algeria, Pakistan, etc.) undertaken by private corporations, or in more specific tasks such as the integral development of iron ore mining and transportation in Algeria, quoted by Barber.[13] In the latter example, the Chase Manhattan Bank, General Electric, and the Chinese People's Republic are reported to co-operate and share the task of planning and co-ordinating the project. Another start may be seen in corporations providing essential educational services to countries in which they are active (e.g. Shell).

Much will depend on the responses business in general will develop to the challenge of becoming what Daniel Bell[18] calls a 'knowledge institution' and regards as one of the important features of the post-industrial society. Taking up this challenge would mean going beyond production and service activities and forming what has already been named the 'quarternary sector', focussing on knowledge and information. Competition, for this sector, would no more take place in terms of products and services, but in terms of organising the latter, in terms of inventing, designing, and planning the systemic structures in which production and services can become effective. Such a trend toward 'social software' as a new task – more, a new role – is already becoming visible in business today. Giving it a world perspective would imply gaining a flexibility, plurality and effectiveness which governments may never attain even if they agree on a global approach.

If business can shift part of its activities to *'global software'*, the global imbalances in the production pattern may have a better chance of being ultimately corrected than with a continuation of present structures and attitudes. 'Harmony and Progress', the motto of the Osaka World Exposition, as interpreted by Saburo Okita – harmony for the developed, and progress for the developing countries – will become a realistic programme only if co-ordinating a pluralistic world approach changes also the course on which Western society is set. For a transitory period, in which 'progress' in production is mandatory for large regions of the world, both approaches to world integration sketched above – an enlightened 'Western' approach of decentralising technological and managerial innovation, and a 'transcultural' approach – may be combined, if the global perspective is

[18] Daniel Bell, 'The Balance of Knowledge and Power', *Technology Review*, June 1969.

maintained. This may become important in developing a certain basic infrastructure which may be viewed as the common prerequisite for many cultural varieties.

Such a combination strategy would also be useful in the development of new technologies that will provide an opportunity to structure development tasks and problem complexes in *new* ways, and introduce them by actually building the new basic structures. A possible example for this task may be seen in the development of the so-called Single Cell Protein (SCP) technology, a group of technologies to produce edible protein in the form of single-cell organisms (bacteria, yeasts, algae) from petroleum, organic waste, or in fresh or salt water tanks or pools. Such a technology would make it possible to set up SCP production in arid or desert zones (e.g. near petroleum reserves, or in arid coastal zones), or in jungle areas (where organic waste abounds). The predominant problem of such regions, namely food, could be substantially alleviated, an industry could be built and an economic basis developed for regions which are considered of low potential so far. The task of planning and developing agro-industrial complexes for such regions (settlement, power generation, water desalination, industry and agriculture, all in an integrated approach), is already taken up by national laboratories in the United States, with contributions from industry; it may also be considered a task for the emerging 'world corporation'.

Two considerations will be of importance: The first is, that there is no World Government to turn to for the creation of a 'public goods' market – in contrast to the national scene where government may be asked to provide, for example, a 'pollution market'. Plans, not geared expressly to goals of specific subsystems, but developed with a world view, will frequently become categorised as 'public goods' for which no government will easily accept any responsibility. The *World Bank Group*, up to now instrumental in implanting 'Western' package deals, probably has the potential to become a trans-cultural institution, too. In this case, it may also become instrumental in stimulating and orienting the development of the 'world corporation' as its chief partner in planning and co-ordinating. It might also become the guarantor of a special trans-national charter for private 'world corporations'.

The second consideration concerns the tentative formation of policies as a necessary prerequisite for the 'world corporation' to clearly define its proper role. Certainly, such policies should not be developed in isolation. Jointly sponsored studies may be envisaged to develop a 'common background' for individual corporate policies, and Herman Kahn and his associates at the Hudson Institute have recently embarked on a forerunner to such studies, sponsored by 60 international corporations and focussing on political, social, economic and technological developments in the coming decades and on a global scale. Other studies may be undertaken for groups of corporations active in a specific functional area. It would, for example, be of utter importance for the food production and distribution industry to know with more precision than anyone can offer today, whether non-agricultural food production technologies will become necessary for the world or not; the short-range or hindsight trends published by FAO can only serve to obscure the long-range picture. Ultimately, the 'world corporation' may be expected to be helped by 'Institutes for the Future' of the type developed at present in Middletown, Connecticut, partly under the patronage of the National Industrial Conference Board – but with global, not national or Western, scope. Such global institutes may, of course, then depend partly on contributions from regional or cultural study centres. In this respect, it is of interest that the Organisation of American States is considering the possibility of setting up a Latin American Institute for the Future.

Such 'Institutes for the Future' may start by investigating the *opportunity costs* for acting in an anticipatory way in areas of imminent crises – and the 'penalty costs' of acting in the wrong way, or not acting at all. A world full of turmoil will also be the death-blow to the international corporation as it thrives today, and as it is setting out to operate in a world market. The active development of roles and institutional structures which are capable of offering effective partnership in the trans-cultural regulation of the world system, will pave the way for the evolution from the multi- and international corporation to the true 'world corporation'.

14. Growth and World Dynamics

A BOOK by Jay W. Forrester[1] extends the application of the *System Dynamics* approach to the largest of all social systems, the world system. The significance of this ambitious step is to be seen not so much in tackling bigger scale and higher complexity – the first world model is, in fact, simpler than the first model for 'Urban Dynamics'[2] and some of the 'Industrial Dynamics'[3] models – but in dealing, for the first time, directly with the ultimate metasystem of which all social systems are part. Since the optimisation of subsystems does not add up to good overall system dynamics and as we are learning in a world of corporate and national 'free enterprise', we may expect radically different policy guidelines to emerge from a 'whole world' approach. Even the first crude results of Forrester's studies bring this out in a very dramatic way: raising agricultural productivity, boosting industrialisation, or applying science and technology to develop replacements for natural resources – all high on the priority list accepted by the world today, hardly ever questioned – now appear as major factors precipitating and aggravating, not (as commonly assumed) staving off, disastrous world crises!

Paradoxically, the nature of complex system behaviour appears to be less counter-intuitive at the world than at a subsystem level. We cannot simulate in our mind the interactions going on in the high-order, multiple-loop, non-linear feedback structures of complex systems. But we can, for the world, recognise absolute and unambiguous system boundaries. We do not have to close off the world system in any arbitrary way – it *is* closed off, if expansion into the hostile environments of space is discounted. We can relate directly to notions such as balance and imbalance, ultimate limitations as well as potentials, and this, in principle, enables us to:

—recognise ultimate implications of current world policy and determine the direction of policy changes which would improve world dynamics in the long run;

[1] Jay W. Forrester, *World Dynamics*, Wright-Allen Press, Cambridge, Massachusetts 1971.
[2] Jay W. Forrester, *Urban Dynamics*, MIT Press, Cambridge, Massachusetts 1969.
[3] Jay W. Forrester, *Industrial Dynamics*, MIT Press, Cambridge, Massachusetts 1961.

—estimate the time-scales involved in the build-up of world crises and the effective introduction of alternative policies;

—calculate the necessary conditions for a steady-state behaviour of parts of the system, and thereby gain a first understanding of the conditions for world balance.

Dynamic modelling would then be necessary and useful primarily to study the possible transitions from growth to balance in the *early* phases of crisis build-up, in order to derive policy guidelines for this most difficult period – the period which we are entering now.

Why, then, is 'direct' thinking of such a penetrating quality still so utterly rare? Why is the thought considered 'unthinkable' that the industrialised societies of the world may have 'overshot' their appropriate levels and now may have, not just to decelerate or stop growth, but to reduce their technology and material consumption levels? Why has Forrester's new book aroused furious and chiefly emotional reactions before it was even published? And what is the nature of that protective mind-block which prevents us from 'thinking the unthinkable', or even from accepting such thinking, and tends to lure us toward the empty promises of 'desirable futures', the false empiricism of linear thinking, and the romantic intoxication with anarchist ideals (of which nationalism and 'free enterprise' are expressions at their appropriate levels)?

Perhaps computer simulations such as Forrester presents them in his book, will convey a clearer and more dramatic message to rationally minded people – as they may be expected to be denounced as inhuman and Orwellian by German 'futurologists' and frustrated social scientists. Computer simulation has a powerful potential of increasing our awareness in areas where human thinking fears to tread, besides promising to become the most important intellectual tool for studying those world policies which appear feasible to effectuate the transition to a more balanced world system.

Is World Dynamics, then, a means for world 'crisis management'? Yes, and the derogatory meaning given to this notion by the advocates of revolutionary 'happenings' implies only a total lack of historical understanding. Policy planning does mean changing the world, and the relations within its system, in a radical way. The choice is between a rational and an irrational revolution, not – as we are made to believe by ideological partisans – between the 19th century concepts of capitalism and socialism. But the rational revolution, for the first time in mankind's evolution, depends on *anticipatory* action, not on mere re-action to crises, because of the speed to which the unfolding of history has been accelerated by man.

Many people, including the bulk of the concerned youth, look at the world from an angle of view of *distribution*, or 'social justice' at world level. This noble attitude remains relatively meaningless if it is not combined with a thorough understanding of the pressures acting on the world system from its boundaries. In other words, if we do not keep asking in a penetrating way 'How big is the world?', we will lack the basis for restructuring it. Both the boundaries and the infrastructure matter for the overall condition of the world system and its dynamics. The logical prerequisite for a world policy is a world loyalty, as recently evoked by one of the leading functionaries of the United Nations.[4] This concept is at present rejected even more furiously by the developing than by the highly industrialised countries. Forrester's studies hit the sorest point in the conventional pattern of thought and action – the predictable outcry against his conclusions will show that.

Forrester approaches his task – this should be emphasised here – in an unpretentious way, with due caution as well as free of illusions. Three quotations from the book clearly mark his personal position: 'The intent is to show that world aggregate variables can be inter-related and that from the interactions can come better understanding of system behaviour.' And again, 'one should not expect models of the kind discussed in this book to predict the exact form and timing of future events. Instead, the models should be used to indicate the direction in which the behaviour would change when changes are made in the system structure and policies.' Also, 'before one is justified in devoting large amounts of time to individual assumptions, he should see something of the total system behaviour'.

The specific cut Forrester takes at the world system in order to model it, brings the interaction between five important *levels* in the world system into focus. These five levels are: population, food, natural resources, industrialisation, and pollution. To simulate their interaction, the model inter-connects concepts from demography, economics, agriculture, technology, and psychology, expressed through fifty-two explicit relationships, governing the rates of change of the five levels through various, and simultaneously acting factors (e.g. changes in the population level from 'normal' birth and death rates, as modified by 'multipliers' from crowding, pollution, food availability, material standard of living, etc.).

Since Malthus' time, all sorts of concern over ultimate limitations in the world system have been voiced – but, and this is the essential point, all of them focussed on *one* kind of limitation only. For Malthus, it was food, periodically it was the depletion of natural resources, and more recently pollution has become recognised as a possible limitation to growth. However, dealing only with one limitation at a time leads to wrong conclusions and policies. The truly *historic significance* of Forrester's approach lies in the simultaneous treatment of a multitude of pressures, each of which may bring the world system to collapse. Four such pressures are taken into account in the first world model: depletion of resources,

---

[4] Philippe de Seynes, Under-Secretary-General for Economic and Social Affairs, in his opening speech at the First Session of the U.N. Committee on Natural Resources, New York, 22 February 1971.

pollution, crowding, and food shortage. And the dramatic message of the book is that *any one of these pressures may be expected to bring the world system down within a few decades, if we fail to introduce world policies in time, designed to cope with all of these pressures simultaneously*. To demonstrate this, the book investigates, *inter alia*, the following cases of no or one-sided policy change:

1. *No change in present world policy*, characterised by net population growth, steady increase in accumulated capital investment (industrialisation), etc.: The pressure coming into play first, in this case, is depletion of natural resources; it brings the world system down gradually, without sharply defined catastrophe (no change in policy may thus be considered not the worst of the alternatives!). Quality of life decreases after the 1950s, population peaks in 2020, the industrialisation trend is reversed around the middle of the next century and consequently pollution, too.

2. *Assumed policy change in 1970 to the effect that natural resource usage is reduced by 75 per cent* through the more effective use of technology (recycling, nuclear energy, replacement of natural materials, etc.): The pressure coming into play first, and in a catastrophical way, is now pollution. Between 2040 and 2060, population drops to one-sixth due to a rapid decline in quality of life.

3. *Assumed policy change in 1970 to the effect that the rate of capital accumulation (industrialisation) is increased by 20 per cent* in an effort to reverse the beginning decline in quality of life: Again, pollution comes into play first, and again in a catastrophical way, but 20 years earlier than in case 2. Quality of life can be maintained at the present level not longer than until the middle of the 1980s.

4. *Assumed policy change in 1970 to the effect that the rate of capital accumulation (industrialisation) is increased by 20 per cent, and simultaneously the 'normal' birth rate (i.e. under present conditions) is reduced by 50 per cent:* Although population stays roughly at the present level, and (in consequence thereof) quality of life rises by 75 per cent to a peak in 2000, pollution again intervenes in a catastrophical way; by 2020, quality of life is down to its 1970 level, and decreases further, along with population, in the following two decades.

5. *Assumed policy change in 1970 to the effect that the 20 per cent increase in the rate of capital accumulation (as in case 3) and the 75 per cent reduction of natural resource usage (as in case 2) are combined:* Quality of life is maintained at about the present level until 2000, but the following pollution catastrophe (due to a particularly rapid and unrestrained increase of industrialisation) intervenes even sooner, and more severely, than in case 3.

6. *Assumed policy change in 1970 to the effect that the 20 per cent increase in the rate of capital accumulation and the 75 per cent reduction of natural resource usage (as assumed in case 5) are now combined with a reduction by 50 per cent of the 'normal' rate of pollution generation* (through application of new technology): Pollution, nevertheless, remains the most important pressure to intervene, a little later (after 2040) and less catastrophical, but hitting now a larger number of people, whose quality of life has already been degraded by crowding.

7. *Assumed policy change in 1970 to the effect that capital investment rate is reduced by 40 per cent, birth rate is reduced by 50 per cent, pollution generation is reduced by 50 per cent, natural resource usage rate is reduced by 75 per cent, and food production is reduced by 20 per cent:* Population, accumulated capital investment, and pollution stay at an equilibrium level (population slightly below its 1970 level), quality of life gradually increases by a total of about one-third, and natural resources continue to be depleted, but at a relatively slow rate which does not affect the equilibrium for about two centuries.

How realistic is this model, and how serious should the results be taken? Clearly, quantitative accuracy is beyond the scope of this first model. It is highly aggregated to the point of assuming, as a first approximation, a homogeneous world. Possibly, different types of system pressures may be expected to come into play first in different regions of the world. However, it should be taken into account that some of the major system pressures – in particular, food shortage – act with shorter time delays than others. In this case, population growth will be restrained in a general way (i.e. people die quickly, if there is nothing to eat), without leading to big fluctuations in the major system levels.

The reported model also contains a rigid set of assumptions, corresponding to a rigid behavioural pattern, which obviously leads to unrealistic results during a major crisis period and afterwards. That quality of life, in line with pre-crisis assumptions, should soar after population has been reduced to one-sixth within two decades, and that this would lead to a rise in population again and to the repetition of all the mistakes committed before the first crisis or catastrophe, would imply the negation of any historical process. Also, crucial parameters, such as quality of life, may have to be understood more deeply before they can be 'trusted' as a decisive factor in the simulation runs. In the first model, quality of life is a composite factor determined by (in this order of importance) food availability, material standard of living, crowding, and pollution. Surely, this cannot be all there is to the quality of life. Also, the high material consumption in World War II obviously assumes part of the responsibility for the high quality of life resulting for the early 1940s. On the other hand, the assumptions relating to the capacity of the natural environment to absorb pollution and to the time constants involved (including 'clogging' of the pollution absorption mechanism), obviously play a crucial role in the build-up of the pollution catastrophes which so often occurs in the simulation runs. There are many possible improvements to be considered in these relationships. More important, however, will be research into what relationships matter and should be built into the model.

Many of the shortcomings of this first model, and of which its author is well aware, are shortcomings of the model, not of the basic methodology. A study group is already at work at the Massachusetts Institute of Technology to disaggregate the model, introduce contingent behaviour patterns, and inter-connect more detailed and more thoroughly investigated sets of assumptions.

However, a major shortcoming of the methodology, as it has been developed so far, may be seen in its incapability to deal with political processes by which basic policy changes may be introduced. It is implicitly assumed, or so it seems, that changes in policy can be made by decree, and that the system would change in the decreed direction without resistance (this very resistance would have the possible effect of leading it in a different direction). At the world level this may not be tenable even for a first approximation. For further methodological development, it appears feasible to attempt a marriage between the System Dynamics approach and the theory of multiechelon (multilevel, multigoal) systems.[5] Such a marriage might then permit the study of possible types of co-ordination of subsystem policies toward a world policy. It is already becoming clear that a rational world policy can emerge only from penetrating cultural and political change, not merely from economic and technological change on which our present strategies are based. Therefore, the nature of the corresponding changes in the cultural and political basis of the different parts of mankind has to be elucidated before policy changes can be modelled in a meaningful way. Forrester seems to be aware of this, when he points to dilemmas which become visible in his results, for example the following one: Pollution threatens to destroy the developed countries, but the correct policy to counter this threat cannot be determined in a clear-cut way. If industrialisation continues, population collapses from pollution, but if industrialisation is stopped, population also collapses, due to the failure of the technical support system for the urbanised society. We even do not know today how to convert the armament fraction of the industrial effort to more peaceful ends, without greatly disturbing the economic system. And nobody yet has a viable proposal for a transition from growth to equilibrium economics.

Nevertheless, even the results from Forrester's first coarse-grain model help to focus our awareness on some essential characteristics of the world system: In the current dynamic situation, most of the essential feedback loops are positive, i.e. self-exciting; negative feedback plays a role only when internal system pressures make themselves felt, in most cases only when a major crisis is already upon us. Population generates the pressures to support growth of population – but supporting that growth leads to more population. Accumulated capital investment engenders more capital investment. High pollution levels slow down the natural pollution absorption mechanisms, which in turn leads to higher pollution levels. And so forth. There is just one important negative feedback loop which would correspond to our present preferences, namely birth rate decreasing with higher material standard of living (which may be a wrong way to interpret

[5] See Chapter 11 of this volume.

this correlation, anyway). Some demographers set all their hope on it. But their thinking is fallacious in so far as they neglect all the other systemic relationships. By all what we know, it is inconceivable that all of mankind develops to a material standard of living which is characteristic of the highly industrialised countries, now, because the world system cannot possibly sustain such growth. In Forrester's estimates, each person in a highly industrialised country already places 10 to 20 times the pollution and natural resource usage load on the world environmental system, compared with a person in a developing country. There is only one conclusion to be drawn from that, namely that the highly industrialised countries will have to *lower* their material standard of living, long before world population can be effectively reduced in a controlled way (and thus permitting higher average standard of living). In other words, our benign development strategy – 'grow and let grow!' – is radically wrong.

There are other important and probably valid lessons to be drawn from Forrester's initial studies: The decades separating us from the end of the century are indeed the decisive ones to act on the basis of choice. Industrialisation *qua* manufacturing products is perhaps the single most dangerous factor in current world dynamics, and it should be questioned in relation to economic growth in advanced countries as well as to development. The shift toward the building of software, especially 'social software', as evoked as an emerging major task for industry in Chapter 13, seems to fall outside this immediate concern. The average world-wide quality of life has reached a peak in the past two decades and can only decline from now on, except for a very precarious planned course into the future which would depend, above all, on population stabilisation. However, a straight birth control programme, on its own, would have only low leverage. Technology can relieve internal system pressures only temporarily, and generates only more pressures; its further development has to be questioned deeply and perhaps it will have to be purposefully de-emphasised. The naive belief, that quantitative growth may continue unchecked and qualitative growth may simply be superimposed,[6] is totally unrealistic – like any linear, non-systemic thinking.

In the present dynamic situation, the following options seem to be open to the world, as this reviewer sees it:

1. *'Let things go'* or *'mechanistic system dynamics'*: The rising system pressures will suppress growth in an uncontrolled way. As Forrester puts it: 'Unless we come to understand and to choose, the social system by its internal processes will choose for us. The "natural" mechanisms for terminating exponential growth appear to be the least desirable. Unless we understand and begin to act soon, we may be overwhelmed by a social and economic system we have created but cannot control.'

[6] U.S. National Goals Research Staff, *Toward Balanced Growth: Quantity with Quality*, Report of the National Goals Research Staff, The White House, Washington, D.C., July 1970.

2. *'Social reason'*:
   - (a) *'Social instinct'* or *'social subconsciousness'*: A basic radical change in the cultural basis (through a revival of instinct or other modes of 'direct' insight). The emerging youth culture may give some hope, in this context, but many of its implications are yet unclear. Another possibility would be the revival or world-wide 'discovery' of cultural traits (e.g. in Asian cultures) which seem to be more in consonance with world dynamics, such as cyclical development instead of linear progress, contemplation instead of competition, learning instead of growth, measure instead of greed.
   - (b) *'Social awareness'* or *'social consciousness'*: The same basic cultural change as in (a) above, but through intellectual understanding, leading to the conscious formation and implementation of rational world policies.

3. *'Social system engineering'*: Restructuring the system in such a way, that internal pressures are deliberately selected and brought into play (or compensated) with the purpose of regulating the system in what we have chosen as a 'good' mode. The precarious balance due to nuclear deterrence, may jump to one's mind here as an actual, if ghastly, example. In Forrester's words: 'no sustainable modes of behaviour are free of pressures and stresses. But many possible modes exist and some are more desirable than others.' This view corresponds to Vickers' notion of freedom as a social artefact, as something to be created, not merely to be preserved.[7]

4. *'Direct control'* over all parts and relationships of the system: The ultimate in dictatorship. But with the corruption of power, where would the wisdom to choose and regulate come from?

It becomes evident that System Dynamics has a most important potential role to play for options 2 (b) and 3 above. The 'Club of Rome', a world-wide gathering of independent intellectuals, sponsors World Dynamics studies primarily to enhance the potential of the 'social awareness' option. But Forrester seems to doubt – and he may be justified – that 'social reason' alone will be able to change the world to a balanced state. He points out that, throughout history, population was always controlled by food availability, i.e. by an internal system pressure, and that always a fraction of the world population was undernourished and lived, so to speak, at the fringe of the sustained system mode. Perhaps, if we better understood what Churchman calls the 'ethics of whole systems' – which evidently runs counter to much of the individual ethics in which we believe by tradition or naive choice – we might acquire the courage to engineer the world system by restructuring what we now consider 'non-negotiables'. But a new political process would be needed to do that. The democratic process of bargaining over 'negotiables' to achieve some incremental steps, obviously cannot do it. Hedonistic constructs and other 'futures fiction' cannot be relevant. There is much evidence, that 'mechanistic system dynamics' will take over – or 'direct control', or both – long before

---

[7] Geoffrey Vickers, *Freedom in a Rocking Boat: Changing Values in an Unstable Society*, Allen Lane, The Penguin Press, London 1970.

'social reason' has had a chance to develop its influence on world dynamics which have been accelerated to doubling times in the order of a decade.

With an unclear potential of 'social reason', 'social system engineering' may be left as our greatest rational chance although we do not yet know how to institutionalise it. System Dynamics is potentially the most important methodological approach to guide us in such a precarious undertaking. 'But to develop the more promising modes', Forrester cautions, 'will require restraint and dedication to a long-range future that man may not be capable of sustaining.'

If we fail to respond to the challenge of reason, there may be just one alternative left, the one which Robert Oppenheimer considered shortly before his death, and in deep despair: Let world dynamics continue its present course and go astray as quickly and as massively as possible, so that Nature gets another chance to build up a new ecology – without man, of course.

SCIENCE, TECHNOLOGY AND HUMAN PURPOSE

15. THE SYSTEM OF SCIENCE, EDUCATION AND INNOVATION
16. TOWARD THE INTER- AND TRANS-DISCIPLINARY UNIVERSITY

Planning and introducing a rational science policy would imply the conscious development of roles for science and technology in the pursuit of human and social purpose. The organisation of science and other forms of human thought toward a purpose may be viewed in the framework of a multiechelon system built by empirical, pragmatic, normative, and purposive system levels. Structured approaches to science, education and innovation may be brought to congruence in such a system.

Key notions for the orientation of science, education and innovation toward an overall purpose – a purpose of mankind – are inter- and trans-disciplinary co-ordination, involving two or all of the system levels. Approaches to normative inter-disciplinarity, linking the pragmatic and normative system levels in the task of social system design, are coming into focus today. The far-reaching implications for higher education are outlined and interpreted in the light of the university's role of enhancing the self-renewal of society.

Both Chapters 15 and 16 are adapted from a paper, first published under the title 'Inter- and Transdisciplinary University: A Systems Approach to Education and Innovation' in *Policy Sciences*, Vol. 1, No. 4 (December 1970). They are reproduced here with the permission of American Elsevier Publishing Company Inc., New York. Chapter 16 is based on concepts the author developed while holding a visiting appointment at the Massachusetts Institute of Technology and elaborated in the report *Integrative Planning for the 'Joint Systems' of Society and Technology – the Emerging Role of the University*, Alfred P. Sloan School of Management, Massachusetts Institute of Technology, Cambridge, Massachusetts, May 1969.

15. The System of Science, Education and Innovation

CURRENT discussions on science policy concepts usually focus either on problems of operating in a given framework of institutions and instrumentalities, or on problems of choice and resource allocation arising from the proliferous activities in science and technology. In other words, tactical and strategic questions are debated before policy – namely the role of science and technology in the pursuit of human and social purpose – has been made explicit. What goes today under the name of science policy, and even 'the new science policy' (also called 'second generation science policy'), hardly ever reaches up to the level of policy at all.

Legal frameworks for the control of technology, comparable to traffic-light regulation of street traffic, would pertain to the operational or tactical level. Technology assessment, a particular task of technological forecasting – or, more appropriately, systemic forecasting[1]– would belong to the strategic level, where, to continue with the traffic image, new road patterns and alternative means of transportation are considered. The restriction to the exploratory part of technological forecasting, as in the approach taken by the U.S. National Academy of Sciences,[2] would constitute a grave mistake. The parallel discussions held by the U.S. National Academy of Engineering[3] recognised this and found it even impossible to proceed in actual forecasting (assessment) tasks without introducing the complementary normative (problem-initiated) approach.

But all this is not yet policy, where notions such as purpose, direction, dynamic systems behaviour, values and norms, institutions, and roles in a purposive process come into focus. When it comes to science and technology, few people dare to address policy issues (although all speak of 'science policy') and a tacit belief in an 'autonomous' development of

[1] See Chapter 5 of this volume.

[2] *Technology: Processes of Assessment and Choice*, Report of the National Academy of Sciences, Committee on Science and Astronautics, U.S. House of Representatives, July 1969.

[3] *A Study of Technology Assessment*, Report of the Committee on Public Engineering Policy, National Academy of Engineering, Committee on Science and Astronautics, U.S. House of Representatives, July 1969.

science and technology reinforces the traditional mechanistic view and hinders the recognition of science and technology as general instances of human activity. A widely shared view holds that science and technology cannot be organised, but their uses can be. In other words, the systemic relationships between man and society on the one hand, and science and technology on the other – or between science and human purpose in general – are again reduced to the conventional, non-systemic notions of 'value-free' science, 'neutral' technology, and 'wise' use of technology.

The currently dominant views of the role of science may best be outlined in terms of the four theories of science planning compared by Harvey Brooks.[4] Three of them negate any feedback from social innovation to science: The Michael Polanyi/Derek Price view of science as an autonomous enterprise and the system of science as an entirely self-regulating community underlies one of these theories. As Brooks observes, 'whereas the Polanyi school (tends) to regard science as a delicate plant, which would wilt if interfered with by society, Price seems to regard science as a vigorous weed that society could not interfere with even if it desired', which leads to different behavioural patterns in the academic community. The second theory, science as a consumer good, or generally an autonomous cultural expression, finds its contemporary protagonists in H. G. Johnson and Stephen Toulmin. It views science as part of the basic purpose of society, and may be regarded as the modern version of the attitude which drew an analogy between science and the arts as the creative domain of the individual, and which dominated through centuries. The third theory, science as a social overhead investment, assumes that science 'underlies *all* the purposes of society, and is therefore to be carried out in an organisational structure which is patterned on the conceptual structure of knowledge'. It links science to social innovation in a non-specific way, but preserves, like the first two theories, a belief in an 'internal ecology of science' which remains unaffected by feedback from social innovation.

These three views of the role of science have been nurtured into veritable myths by partisans of a misinterpreted 'academic freedom', who are even stronger in Europe than in America, and who have prompted the Russian physicist Artsimovich to remark during a visit to the Massachusetts Institute of Technology: 'If I understand correctly, in the Western countries just like in the Soviet Union, a scientist is called somebody who is permitted to pursue his hobby at the expense of society.'

Only the fourth theory, Alvin Weinberg's view of science as a technical overhead on social goals, by focussing on the achievement of certain goals through technology establishes a direct relevance of social goals to scientific activity. John Platt[5] has brought this view dramatically into the fore-

---

[4] Harvey Brooks, 'Can Science Be Planned?', in *Problems of Science Policy*, OECD, Paris 1968.

[5] John Platt, 'What We Must Do', *Science*, No. 166 (1969).

ground of discussion. It hardly conceives science and society as aspects of a comprehensive system, but looks at direct relationships between goals of a strategic character and technological contributions toward them. It would stimulate a powerful normative extension of the technology assessment procedures considered by the two Academies, as mentioned above. A reorganisation of scientific activity into normative, though fragmented, inter-disciplinary approaches is the result aimed at. A certain danger may be seen here in the temptation to take a straightforward (non-systemic) problem-solving approach of the type which has proved so successful in attaining purely technological targets, and to neglect the systemic character of most of these problems in the social domain.

For our discussion, the crucial question is whether science and its internal system of relationships is independent from human and social purpose, or whether there is a feedback link tieing them together. We have learned part of the answer by recognising that not only scientific facts, but also scientific structures can be grasped by the human mind only through what we may call *anthropomorphic modes of organisation* – and we have further learnt that these modes are neither isomorphic nor even unambiguous when applied to the structures of reality as we understand them. Modern physics is basically about the creation of anthropomorphic models of an 'inhuman' structure of reality.

The other part of the answer, concerning feedback from social organisation and dynamics to science, is finding its tentative formulation today with the emergence of 'technological man' (Ferkiss) and the first traits of a post-industrial society in the face of growing complexity and uncertainty and a seemingly amorphous, disquieting world *problématique* (Ozbekhan). Until recently, in the above scheme until Weinberg's propositions, such a feedback has rarely been explicitly recognised in theories of science and science planning.

Attempts to elaborate on Karl Mannheim's sociological approach to science, his 'Wissenssoziologie', and to conceive science in the framework of a 'social construction of reality',[6] generally take a phenomenological approach today which fails to perceive a dynamic and purposive science/innovation system. The same holds for Lévi-Strauss' observation of a certain parallelism in the structures of science and societal behaviour. Most of these attempts try to preserve science as a 'value-free' abstraction. The spearheads of a critical sociology (e.g. H. Marcuse and Habermas) which do recognise the dynamics of social innovation, and the role of science in human activity, frequently disregard the full human potential of purposeful design of the social reality through the overall science/innovation system.

---

[6] See, for example, Peter Berger and Thomas Luckmann, *The Social Construction of Reality*, Allen Lane, The Penguin Press, London 1967, and Penguin University Books, Harmondsworth 1971.

A systems approach, as it is proposed in this chapter, would consider science, education and innovation, above all, as general instances of purposeful human activity, whose dynamic interactions have come to exert a dominant influence on the development of society and its environment. *Knowledge* would be viewed here as *a way of doing, 'a certain way of management of affairs'* (Churchman). Among other things, a new policy as well as new structures for the knowledge institutions, such as the university, may be expected to emerge from such an approach. They will constitute responses to the *specific situation* in which society and science find themselves today, and will be subject to continuous change. As a matter of fact, they ought to be designed explicitly with a view toward their innate capability for flexible change in accordance with the dynamically evolving situation – in which science and technology may not always play the role they play today.

Thus, we find ourselves in a dilemma, from what angle of view (what policy) to try to elucidate the structures of science – God's or man's? What is man's principal task in dealing with science, perception or creation? Do we want Mozart or Beethoven – the image of man in God, or that of God in man –, Palestrina or Wagner – meaningful structure or structured, passionately individual meaning? Is choice really a necessity here? Is there a choice at all?

The answer is that there is no resolution to this dilemma, that it constitutes perhaps one of the basic paradoxes with which we must learn to live, and which give enhanced meaning to human life. The *human condition* in the scientific-technical era may, again, find its ultimate expression in the spirit of the ancient Greek tragedy, in which man attains his full creative freedom in making the 'structures' (not simply the laws) imposed by the Gods his own for taking purposive action.

Organisation for a purpose implies the introduction of normative and pragmatic principles which are beyond the traditional notion of empirical and empirical/conceptual science. Which part of the science/innovation system is accepted under the name of science, and which is not, is totally unimportant. What matters is that science is recognised as part of the human and social organisation.

The task is nothing less than to build a new society and new institutions for it. With technology having become the most powerful change agent in our society, decisive battles will be won or lost by the measure of how seriously we take the challenge of restructuring the *'joint systems' of society and technology* – the systems of which both society and technology are integral constituents, systems of urban living, environmental control and conservation, communication and transportation, education and health, information and automation, and so forth. And the outcome of these battles will depend, above all, on the competence and imagination of people in the key institutions dealing at present with science and technology: Government at all jurisdictional levels, industry and the university. They have, in the recent past, acquired some skill in inventing, planning and designing complex technical systems. More than anything else, our propensity for actively shaping our future will depend on the extent to which, and on the pace at which, these key institutions – or entirely new types of institutions replacing them – will acquire the capability to deal effectively with systems in an integrative way, cutting across social, economic, political, technological, psychological, anthropological and other dimensions. These systems will embrace Nature, man, society, and technology.[7] Applying to them sequential measures of control and problem-solving can only aggravate the situation from which so clear a lesson can be learned.

We may now try to tentatively design an integral science/innovation system, or even a *science/education/innovation system* (if education is accepted as being essentially education for the self-renewal of society), in which science, education and innovation become aspects of one and the same structure of thought and action. Such a system would constitute a most suggestive example for the systems notion according to a recent definition:[8] A system is a relationship among objects described (or specified, defined) in terms of information processing and decision making concepts.

Scientific disciplines become organised in such a system in a particular way which depends on the *normative* orientation of science, education and innovation. The boundaries of disciplines, their interfaces and inter-relationships no more correspond to an *a priori* system of science. This is the *human action model* approach, as distinct from approaches based on a mechanistic model, with which man does not interfere.

Figure 15.1 attempts to sketch a possible way of organisation in the form of a multiechelon hierarchical system.[9] The angle of view applied here is that of the improvement of the system of human society and its environment – a *partisan* angle of view which starts from the assumption that man has become the chief actor in the process of shaping and con-trolling the system. It may be called the *anthropomorphic angle of view* which, by definition, cannot be 'objective'. Nor would it be possible to form the notion of an integral science/education/innovation system with-out a purposeful, thus also dynamic, and inherently 'subjective' view in mind.

The traditional dissection of knowledge and knowledge transfer into a variety of disciplines has been developed from another angle of view, namely that it should be possible to arrive at a mechanistic explanation of the world *as it is* by putting empirical observation into a logical context.

[7] See Chapter 3 of this volume.

[8] Mihajlo D. Mesarović, *Systems Concepts*, paper prepared for the UNESCO Project 'Scientific Thought', 30 November 1969, revised version 29 December 1969.

[9] See also Chapter 11 of this volume.

Such science has been called by Churchman 'simply a defective part of the social organisation'.[10] Disciplinarity in science is essentially a static principle which becomes meaningless if considered in the framework of a purposeful system. It is no wonder that in a time when science becomes increasingly understood as basis for, or even integral aspect of, creative human action, the emphasis shifts to more or less inter-disciplinary approaches. However, it has not yet become fully clear what inter-disciplinarity and the intermediate steps toward it really mean.

In a purposeful science/innovation system, inter-disciplinarity has to be understood as a *teleological* and *normative concept*. Above all, we must ask: Inter-disciplinarity to what end? It involves the organisation of science toward an end, in other words the linking of adjacent hierarchical levels in the system as sketched in *Figure 15.1*, with the aim of co-ordination.

With this notion, and with the introduction of a purpose, the science/innovation system according to *Figure 15.1* assumes a specific meaning in system theoretical terms. Otherwise, it might at first look appear as a stratified system, where the different strata signify levels of abstraction. Each stratum would then have its own set of terms, concepts and principles. Crossing the strata in the downward direction would give increased detailed explanation, while crossing them in the upward direction would give increasing significance. Empirical science, in many instances, has been developed in such a stratified way. The example of the biological sciences with their strata from whole organisms and organs down to cells and further to molecules provides here the most suggestive example.

In a purposeful system, or human action model, however, inter-disciplinarity constitutes an *organisational principle* for a two-level co-ordination of terms, concepts and disciplinary configurations which is characteristic of a *two-level multigoal system*.[11] The important notion here is that with the introduction of inter-disciplinary links between organisational levels the scientific disciplines defined at these levels change in their concepts, structures, and aims. They become co-ordinated through a *common axiomatics* – a common viewpoint or purpose.

It should be noted here that the four hierarchical levels depicted in *Figure 15.1* are further subdivided into a fine-structure of hierarchical sublevels. For example, there are such levels between basic technologies and complex technological systems, between relativistic physics and macro-theories such as reactor core physics. The notion of inter-disciplinarity may also be applied to links between these sublevels, which

[10] C. West Churchman, *Challenge to Reason*, McGraw-Hill, New York 1968.

[11] For the different concepts of hierarchical systems, see: M. D. Mesarović and D. Macko, 'Foundations for a Scientific Theory of Hierarchical Systems', in L. Whyte *et al.* (eds.), *Hierarchical Structure*, Elsevier, New York 1969. For a rigorous treatment of multilevel systems, see: M. D. Mesarović, D. Macko, and Y. Takahara, *Theory of Hierarchical Multi-Level Systems*, Academic Press, New York 1970. See also Chapter 11 of this volume.

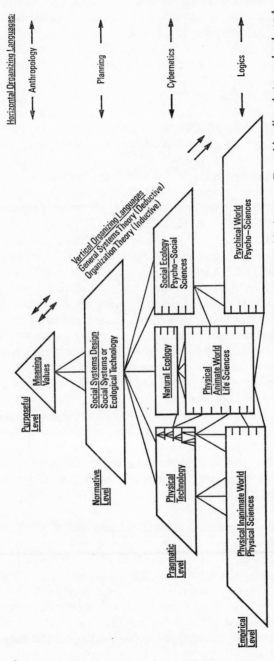

Fig 15.1 The Science/Education/Innovation System, viewed as a multiechelon hierarchical system. Branching lines between levels and sublevels indicate possible forms of inter-disciplinary co-ordination.

may be formed across different 'blocks' of science, for example in bio-chemistry. The essential is that a new common axiomatics can be intro-duced from the higher level.

The ultimate degree of co-ordination in the science/innovation system, finally, which may be called *trans-disciplinarity*, would not only depend on a common axiomatics – derived from co-ordination toward an 'overall system purpose' – but also on the mutual enhancement of epistemologies in certain areas, on what Ozbekhan[12] calls 'synepistemic' co-operation. With trans-disciplinarity, the whole science/innovation system would be co-ordinated as a *multiechelon* (*multilevel, multigoal*) *system*, embracing a multitude of co-ordinated inter-disciplinary two-level systems, which, of course, will be modified in the trans-disciplinary framework.

Trans-disciplinary concepts and principles over the whole system change significantly with changes in the 'overall system purpose' toward which the top co-ordination function of 'meaning' in *Figure 15.1* is oriented. We lack a deeper understanding of this purpose, and thus an unambiguous direction for our organisational efforts. Nevertheless, we cannot hope to act with a true purpose – in other words, to manage the multiechelon science/innovation system in a meaningful way – if we do not search for and bring into play values and norms, a *policy for mankind*,[13] to guide the development of science and technology, education and innova-tion. Our best efforts must therefore focus on the top structure of the system. Only through such efforts may we hope to arrive at a real science policy.

The various steps of co-operation and co-ordination among disciplines, as they are currently discussed with a view to higher education, may now be defined and, at the same time, identified as organising principles for hierarchical systems of increasing complexity, as proposed in *Figure 15.2*. It was necessary to introduce here a new intermediate step, which is tentatively called cross-disciplinarity and which threatens to blur aims and purposes in the development toward higher forms of co-ordination. As a matter of fact, most of the approaches called 'inter-disciplinary' today are at best pluri- or cross-disciplinary!

One of the most conspicuous attempts of cross-disciplinary polarisation is the reformulation of management, planning, and organisation – even explicitly the planning of change – in terms of the empirical and reduc-tionist concepts of the applied behavioural sciences. Other cross-disciplinary attempts to dominate by imposing specific disciplinary con-cepts across a level of the science/innovation system, start from economics. Purely economic criteria and linear methods (e.g. econometrics) are applied to scientific research and development, to education, and now also to

[12] Hasan Ozbekhan, *The Future of Education*, paper presented to a meeting of the Frensham Group, March 1970.
[13] See also Chapter 14 of this volume.

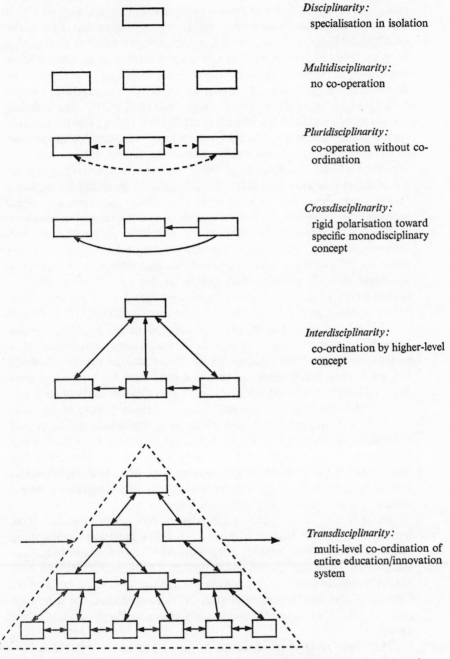

Disciplinarity:
specialisation in isolation

Multidisciplinarity:
no co-operation

Pluridisciplinarity:
co-operation without co-ordination

Crossdisciplinarity:
rigid polarisation toward specific monodisciplinary concept

Interdisciplinarity:
co-ordination by higher-level concept

Transdisciplinarity:
multi-level co-ordination of entire education/innovation system

Fig 15.2 Steps Toward Increasing Co-operation and Co-ordination in the Science/Education/Innovation System.

environmental problems and aspects of socio-technological systems which will be subjected to a crude economy/diseconomy approach. The recent and drastic failure to explain, or at least describe, the 'technological gap', a truly systemic phenomenon,[14] in disciplinary terms – as an economic or trade gap, a market gap, a licence and royalties gap, a technological development gap, a management gap, an education gap, etc. – seems already forgotten. The belief of economists (and partly also political scientists) in the supremacy of their disciplinary thinking, and the easiness with which their claim is accepted in a materialistic world, constitute one of the main obstacles to a systems approach to science and innovation.

Multi- and pluri-disciplinarity involve only the purposeless or purposeful grouping of rigid disciplinary 'modules'. Cross-disciplinarity implies a 'brute force' approach to reinterpret disciplinary concepts and goals (axiomatics) in the light of one specific disciplinary goal and to impose a rigid polarisation across disciplines at the same level. Only with inter- and trans-disciplinarity the science/innovation system becomes 'alive' in the sense that disciplinary contents, structures and interfaces change continuously through co-ordination geared to the pursuit of a common system purpose. *Inter- and trans-disciplinarity thus become the key notion for a systems approach to science, education and innovation.*

The science/innovation system, as sketched in *Figure 15.1*, is built from the bottom level upwards. This is inevitable since, in a multiechelon system, the upper organisational levels cannot achieve anything without the activities at the lower levels, just as a conductor cannot achieve anything without an orchestra. On the other hand, this means that two major obstacles on the way to inter- and trans-disciplinarity have to be overcome: One is the rigidity of disciplines and disciplinary concepts and axiomatics developed at the lower levels; the other one is the application of lower-level concepts and axiomatics to higher levels. Both obstacles, indeed, prove very severe in the development of a meaningful social science and in current approaches toward an inter-disciplinary social technology, as will be briefly discussed below.

At each level, an 'organising language' is tentatively identified. This notion goes beyond that of a merely expressive language or deductive science. An 'organising language' has the quality of an operator in achieving systemic co-operation and co-ordination. Mathematics seems to be a more ubiquitous operator, underlying part of the 'organising languages'. Certainly, there are 'structures' implied in these operators, but it may be more appropriate to consider them as anthropomorphic 'structures', than as 'objective' ones.

The *empirical level* in *Figure 15.1*, with *logics* as its 'organising language', may be subdivided into three bodies of science which all developed

---

[14] The systemic nature of the 'technological gap' has been grasped in the book: Aurelio Peccei, *The Chasm Ahead*, Macmillan, New York 1969.

on the basis of empirical observation and logical interpretation: (a) Physical sciences, with the traditional disciplines; (b) Life sciences, which occupy a special position and extend over both empirical and pragmatic levels, from basic knowledge up to complex biological systems and parts of medical technology; and (c) Psycho-sciences, which include psychology and the behavioural sciences as well as aspects of human perception and creative expression, such as the arts and religions. These sciences aim at describing the world as it is and at being 'objective', a concept which is at least doubtful in the domain of the psycho-sciences. Inter-disciplinary types of teleological co-ordination have become fruitful particularly between hierarchical levels within the physical sciences as well as between physical and life sciences (e.g. biochemistry on the one hand, molecular biology on the other) and, to some extent, between the life sciences and the psycho-sciences.

The *pragmatic level* with *cybernetics*, the science of regulation and control, as common 'organising language', represents a higher level of organisation and may be subdivided into: (a) Physical technology, embracing many hierarchical sublevels from basic technology over simple products to complex technological systems together with their functional interactions with societal systems; (b) The more systemic part of the life sciences, and natural ecology, which has been successfully used to develop agriculture; and (c) Social ecology, or simply culture, based on psycho-social sciences, comprising, *inter alia*, history, sociology, linguistics and communication in general, communicative aspects of the arts, micro-economics, political science (in its narrow pragmatic meaning), cultural aspects of anthropology, and the traditional ethics of the individual. Or, rather, there ought to be such a science of social ecology, applicable in a pragmatic way.

One of the two obstacles mentioned above prevented so far the full establishment, in an inter-disciplinary way, of the pragmatic level. The narrowly interpreted 'scientific method', or the part of it which is used to defend empiricism and logical positivism, was transferred to the prag-matic level. Physical technology, in many instances, first developed on the basis of empirical observation and logical interpretation, followed by conceptualisation of operating principles such as found in the steam engine, the steam turbine, aircraft, to mention just a few. But all these technologies quickly became inter-disciplinary melting pots for various physical sciences when the need for manipulability, and therefore for theory, arose. To what extent the technology-oriented axiomatics 'trimmed' the concepts of physical science is demonstrated by chemical engineering, reactor physics, or aircraft and rocket design, where complex interactions of microphenomena are cast into handy macrophenomenal theories which suit the needs of specific pragmatic uses of technology to perfection.

Such a swift adaptation did not take place in the area of social ecology, or the psycho-social sciences. Here is the profound reason for the frequently denounced lagging behind of the social sciences. As Churchman[15] remarks, 'perhaps one of the most ridiculous manifestations of the disciplines of modern science has been the creation of the so-called social sciences', which pursue the same mechanistic ideal of 'objective' empiricism as the physical science disciplines. 'Instead of social science partitioning itself into special disciplines, it should recognise that social science is not a science at all unless it becomes a natural part of the activities of social man.' Above all, social science ought to express the potentials of human freedom, creativity, and responsibility. Instead, social science, particularly in the United States, is becoming infested with the reductionist concepts of the behavioural sciences. Neither the old analytical nor the younger phenomenological schools of social science tell us how to conduct our social life, but tend to discourage us from developing any pragmatic or normative, i.e. value-dependent, social science by making us believe that social science is inherently data-rich and theory-poor. The vigorous development of a critical sociology has increased our understanding of the inter-relationships between technology and social science, but does not yet provide the building blocks for a normative social science.

The *normative level*, with *planning* as its 'organising language', deals with social systems design, bringing into focus social systems or ecological technology in its broadest meaning. It has as its core Churchman's 'ethics of whole systems' and branches out into aspects of social systems technology such as law, macroeconomics, and institutional innovation. Typically, it focusses on large social and man/environment systems, ekistics, and a variety of 'joint systems' of society and technology.[16] Few of the domains at this level have yet found valid frameworks – ekistics may be farthest advanced in this respect – and the current concepts of law and macroeconomics hardly meet the inter-disciplinary challenge posed to them in the technological era. It is at this level that the broad conceptualisation of man's active role in shaping his own and the planet's future unfolds.

The *purposeful level*, finally, brings values and value dynamics into play through interactive fields such as philosophy, arts and religions, structuring in an inter-disciplinary way some of the domains at the normative level. The 'organising language' at this level ought to be *anthropology* in its most profound notion, the science of how to create an anthropomorphic world and how mankind may become capable of surviving dynamically changing environments. That most of anthropology today is not much more than an empirical behavioural science, illuminates drastically the confusion created in modern science by the traditional cultural postulate of 'knowledge *per se*' and the corresponding emphasis on empiricism. Of course, the psycho and psycho-social sciences will have to provide impor-

[15] C. West Churchman, *Challenge to Reason*, McGraw-Hill, New York 1968, p. 85.
[16] See Chapter 3 of this volume.

tant fundaments for the new anthropology through a succession of inter-disciplinary 'elevations' of their concepts.

It appears futile to discuss what, in the full science/education/innova-tion system, as sketched in *Figure 15.1*, should be called science and what not. In a narrow, positivistic sense, the notion of science applies only to the lowest systems level. Whether this science is organised and co-ordinated again by science or by categories of thought, which are given other names, is a matter of arbitrary definition.

The horizontal 'organising languages' of logic, cybernetics, planning and anthropology, in order of increasing systemicity, intermesh with the vertical 'organising languages' of *general systems theory* (deductive) and *organisation theory* (inductive). If the science/innovation system is viewed as a purposeful system for the self-renewal of society, as outlined above, we should, in Ozbekhan's words,[17] 'be able to investigate in a more orderly way than has hitherto been possible, whether methodologies arising from anthropology and general systems theory – both of which deal with phenomena that pertain to whole groups – might not be forged into a methodological structure for planning'. With such a structure for planning it will then be possible to link the normative, pragmatic and empirical levels in an inter-disciplinary way and ultimately aim at a genuine trans-disciplinary co-ordination, i.e. at managing and science/education/innovation system in an integral way.

Then only we may also hope to develop concepts of a *rational science policy*.

---

[17] Hasan Ozbekhan, 'On Some of the Fundamental Problems in Planning', *Techno-logical Forecasting*, Vol. 1, No. 3 (March 1970).

16. Toward the Inter- and
Trans-disciplinary University

EDUCATION FOR SELF-RENEWAL

WE are living in a world of change, voluntary change as well as change brought about by mounting pressures outside our control. Gradually, we are learning to distinguish between them. We engineer change voluntarily by pursuing growth targets along lines of policy and action which tend to rigidify and thereby preserve the structures inherent in our social systems and their institutions. We do not, in general, really try to change the systems themselves. However, the very nature of our conservative, linear action for change puts increasing pressure for structural change on the systems, and in particular, on institutional patterns.

We are baffled by the sudden appearance of such pressures for change in the educational system by student unrest and by the notion that the current type of education may no longer be relevant. We are confused by the degrading side-effects of technology on the systems of human living, in the cities as well as within the natural environment. And we are ridden with doubts about the effectiveness of decision making processes dominated by short-range and linear thinking and about the piecemeal and passive way in which scientists and engineers respond to them. Through its three functions – education, research, and service – the university is deeply affected by all of these pressures for change. To live with them, to absorb them and even make use of them, requires a new purpose and a new structure for the university.

Looking at changes, and pressures for change, in all three of the primary functions at the university, we may, *inter alia*, discern the following important trends:

—*Education:* From training for well-defined, single-track careers and professions (by duplicating existing skills) toward an education which enables judgment of complex and dynamically changing situations – in other words, geared to the continuous self-renewal of human capabilities, with emphasis shifting from 'know-how' to 'know-what';

—*Research:* From discipline-oriented research over pluri- and inter-disciplinary research toward research on complex dynamic systems – or, from research on the fundamental level and the perfection of specific technologies to the organisation of society and technology in a systems context;

—*Service:* From specialised, piecemeal research contributions and passive consultations to an active role in the planning for society, in particular, in the planning of science and technology in the service of society.

A synopsis of the pressures for change, as recognised above for the individual functions of the university, and those for change in society at large, yields a picture of powerful forces which act disruptively within the existing structures, but seem to converge reasonably well in their ultimate meanings and implications. The new purpose of the university may readily be found in this area of convergence of reason. It may be expressed as the new purpose of the institution itself, not of its members.

In most general terms, the purpose of the university may be seen in the decisive role it plays in *enhancing society's capability for continuous self-renewal.* It may be broken down further in line with the principal characteristics of a society having this capability, as spelled out by John Gardner:[1]

Enhancing the pluralism of society, by bringing the creative energies of the scientific and technological community as well as of the young people, the students, fully into play – not for problem-solving, but for contributing to society's self-renewal;

Improving internal communication among society's constituents by translating into each other the mutual implications of science and technology on the one side, and social objectives on the other, and by pointing out the long-range outcomes of alternative courses of action in the context of broadly conceived social systems;

Providing positive leadership by working out measures of common objectives, setting priorities, and keeping hope alive, as well as by promoting experiments in society through ideas and plans, and, above all, by educating leaders for society.

The new purpose implies that the university has to become a *political institution* in the broadest sense, interacting with government (at all jurisdictional levels) and industry in the planning and design of society's systems, and in particular in controlling the outcomes of the introduction of technology into these systems. The university must engage itself in this task as an institution, not through the individual members of its community.

The university ought to become society's strategic centre for investigating the boundaries and elements of the recognised as well as the emerging 'joint systems' of society and technology, and for working out alternative

[1] John W. Gardner, *Self-Renewal: The Individual and The Innovative Society,* Harper and Row, New York 1965; and Godkin Lectures, Harvard University, March 1969.

propositions for planning aimed at the healthy and dynamically stable design of such systems.

The major changes which this new purpose will bring to the university, include the following ones:

Principal orientation toward socio-technological systems design and engineering at a high level, leading to emphasis on general organising principles and methods rather than specialised knowledge, both in education and research;

Emphasis on purposeful work by the students rather than on training;

Organisation by outcome-oriented categories rather than by inputs of science and technology, and emphasis on long-range outcomes.

With the new purpose, the education, research and service functions of the university, which have increasingly come apart, will again merge and, in fact, become one. This emerging unity corresponds to the integral view of the education/innovation system elaborated in Chapter 15.

CURRENT APPROACHES TOWARD NORMATIVE INTER-DISCIPLINARITY IN THE UNIVERSITY

How far has the university gone in penetrating the education/innovation system? Clearly not very far yet. In particular, the education function of the university was not capable of adjusting to the requirements of inter-disciplinary organisation beyond the level of elementary technology. To a large extent, education in technology is still categorised by disciplines and departments called 'mechanical engineering', 'electrical engineering', 'chemistry', etc. This has led to two grave consequences: One is a schism between the education and research functions of the university at levels of higher inter-disciplinary organisation, which is becoming a problem already at the level of complex technical systems; university research and development in these areas is increasingly set up and carried out outside the educational structures. The other consequence is a growing mismatch between engineering education and the requirements of industry which is reorganising itself in terms of technological or even socio-technological systems tasks.[2] In the contemporary Institute of Technology, computers and information technology still are subsumed under 'electrical engineering'; no wonder, then, that the particularly strong systemic interaction between man and computer has not yet found a place in the university.

The sore state of social science will not improve rapidly where conventional social science departments deal with the conventional wisdom of empirical or behavioural social science. However, innovative university programmes, particularly geared to undergraduate study, are paving the

[2] See Chapter 8 of this volume.

way to a meaningful pragmatic and normative social science. Theme Colleges of Community Science and of Creative Communication at the Green Bay campus of the University of Wisconsin provide a good example here. Even more significant may become the influence of systems-oriented educational research programmes, such as Urban, Regional, and Environmental Centres or Departments, which may be expected to create their own approaches to social science if what is readily available has to be judged irrelevant for social systems design.

In the meantime, the social side of the education/innovation system produces a number of cross-disciplinary approaches which all have in common that they fail to recognise the systemic character of science and technology as integral aspects of the 'joint system' of society and technology. One of the most conspicuous attempts of cross-disciplinary polarisation is the reformulation of management, planning, and organisation – even explicitly the planning of change[3]– in terms of the empirical and reductionist concepts of the applied behavioural sciences. Other cross-disciplinary attempts to dominate a level of the education/innovation system by imposing narrow disciplinary concepts, start from economics. The Organisation for Economic Co-operation and Development (OECD), for example, has applied purely economic criteria and linear methods (econometrics) to education and to scientific research and development, and is setting out to do the same to environmental problems and aspects of socio-technological systems (through a crude economy/diseconomy approach).[4] The recent and drastic failure to explain, or at least describe, the 'technological gap', a truly systemic phenomenon,[5] in disciplinary terms – as an economic or trade gap, a market gap, a licence and royalties gap, a technological development gap, a management gap, an educational gap, etc. – seems already forgotten. The belief of economists in the supremacy of their thinking, and the easiness with which their absolute claim is accepted today, constitute one of the main obstacles to a systems approach to education and innovation and to the development of inter- and trans-disciplinary organisation in the education/innovation system.

Most of the current university experiments emerging from the social side of the system and expressing themselves in Schools of Public Affairs, Public Policy Programmes, or Programmes in Policy Sciences, constitute essentially *cross-disciplinary approaches*. Instead of one traditional discipline, a group of 'soft' disciplines may dominate and provide concepts, principles, and methods which are applied to the entire pragmatic level of the education/innovation system. A good example is Harvard's Program

[3] See, for example, the anthology: Warren G. Bennis, Kenneth D. Benne, and Robert Chin (eds.), *The Planning of Change*, Second Edition, Holt, Rinehart & Winston, New York 1969.
[4] OECD Ministerial Meeting, Paris, 20–22 May 1970; press reports.
[5] The systemic nature of the 'technological gap' has been grasped in the book: Aurelio Peccei, *The Chasm Ahead*, Macmillan, New York 1969.

of Graduate Education for Public Service which started in the fall of 1969. It is structured into the four main areas of analytical methods, economic theory, statistical methods and political analysis, whereby existing 'modules' of concepts and methods (mainly pertaining to economics) are employed; not the contents, only the combination is supposed to be new. The implicit assumption here is that a rationale can be found, to which the 'hard' sciences and technology may be subjected without being part of it. In other words, science and technology are seen as 'neutral' tools which may be put to any use, implying also an unbroken faith in sequential problem-solving, i.e. a non-systemic approach. The 'seamless web'[6] into which human society has been transformed by technology, cannot be grasped in this way.

The first steps toward a *normative inter-disciplinary approach* in the university, i.e. a link between the pragmatic and normative levels in *Figure 15.1*, are taken where basic themes of society or need-areas are recognised and accepted for a fundamental reorganisation of the educational and research disciplines involved. The scientific-technical and the psycho-social sides of the education/innovation system become integrated in this approach. It is quite obvious that only universities with well-developed structures on both sides can take this approach. The discussion whether universities should deal with technology, or Institutes of Technology should adopt social science – a discussion which, in Europe, is still dominated by a belief in a fundamental polarisation between scientific-technical and humanistic cultures (C. P. Snow's 'two cultures') – finds its resolution in the normative systems approach. On the other hand, it still presents a hard-to-overcome obstacle on the university's way to inter-disciplinarity reaching up to the normative level.

Some university structures, corresponding to this approach, focus on the *educational* function. Significant large-scale examples are:

The College of Agricultural and Environmental Sciences at the Davis campus of the University of California, organised in five broad areas of systemic nature, including a systems approach to environmental problems;

The Theme Colleges of Environmental Sciences, Human Biology, Community Sciences, and Creative Communication at the Green Bay campus of the University of Wisconsin, currently geared to undergraduate education, with graduate programmes in preparation;

The Program in Environmental Science and Engineering at the School of Engineering and Applied Sciences of Columbia University;

The planned Graduate College on the Human Environment at the Madison campus of the University of Wisconsin;

[6] Victor C. Ferkiss, *Technological Man: The Myth and the Reality*, George Braziller, New York, and Heinemann, London, 1969.

A University of Planning (or Environmental Design) at Solothurn (Switzerland), currently in a preparatory stage.

Other structures focus mainly on *research* and frequently assume the
form of inter-disciplinary centres in which faculty members and graduate
students, pursuing their 'formal' careers in traditional departments, may
find a 'second home'. Examples are various urban centres, the Harvard
Program on Technology and Science, the Center for Research on Utilisation of Scientific Knowledge at the University of Michigan at Ann Arbor,
the Center for the Study of Science in Human Affairs at Columbia University, and the Program of Policy Studies in Science and Technology at
George Washington University. A special research domain (*Sonder-
forschungsbereich*) 'Planning and Organisation of Socio-technological
Systems' has been proposed in the Federal Republic of Germany, to be
established at one or two universities selected for a 'focal' approach (most
likely at the Technical University, Hanover). The weakness of many of
these centres lies in their passive attitudes, which do not attempt to
organise and stimulate research on systemic problems to the degree
necessary in view of the complex and inter-disciplinary character of such
research. To some extent, a certain dominance or even cross-disciplinary
claim from the social side may be observed to sneak in, somewhat distort-
ing the original aim.

Of the greatest significance among the steps taken toward normative
inter-disciplinarity are experimental university programmes attempting an
integrated education/research/service approach. Engineering departments
of a conventional type may engage in 'technology assessment' (i.e. technological forecasting in a social systems context), as has been done at the
University of California at Los Angeles (UCLA). To some extent, Schools
or Departments of Architecture, Urban and Regional Planning, or
Environmental Design, have always been explicitly or implicitly systems-
minded, and have developed half-way toward normative inter-disciplinarity
dealing with important areas of social technology. The Athens Centre of
Ekistics (Greece), with its international mixture of students, may serve here
as a small, but stimulating model of a truly inter-disciplinary education/
research approach involving the normative level. More broadly oriented
experiments include the following ones:

Specific socio-technological systems design studies in the framework of the
'Special Studies in Systems Engineering' programme at the Massachusetts Institute of Technology; Project Metran (an integrated
urban transportation system) and the Glideway System Concept
(a high-speed inter-urban transportation system) made considerable
impact by stimulating thinking and concrete systems and hardware
developments, the latter for example in relation to MIT's Project

Transport for a high-speed ground transportation system for the American Northeast Corridor, which became the core of a large decentralised project on a national basis;

The Program on Science, Technology and Society at Cornell University;

The Program in Environmental Systems Engineering at the University of Pittsburgh;

The graduate Program for the Social Application of Technology (PSAT) at the Massachusetts Institute of Technology, which started the fall of 1971;

The planned Center for Advanced Studies – a Systems Center, an Environmental Center, and an Energy Transformation Center, which may well merge into one – at the Hartford Graduate Center of the Rensselaer Polytechnic Institute.

These experimental structures usually have their own faculty and are started with a view to becoming the core for larger innovative structures. They already include many elements of the function-oriented departments, and to some extent also of the systems design laboratories, proposed in the following section.

On the other hand, the grandiose idea for an international post-graduate 'systems university' to be located in Europe – a concept developed by an international committee and subsequently partly through OECD – collapsed because of lack of imagination in the moment when governments, and through them industrial confederations, became involved. The resulting International Institute for the Management of Technology (IIMT) at Milan (Italy) has now been set up by European governments on the basis of providing a framework for six-week training courses for industrial and public managers. . . .

A TRANS-DISCIPLINARY STRUCTURE FOR THE UNIVERSITY

The essential characteristic of a trans-disciplinary approach is the co-ordination of activities at *all* levels of the education/innovation system. Even the more imaginative proposals for new structures and curricular patterns in the university usually stop short of conceiving a co-ordinated system. Harold Linstone[7] proposes educational strata of pluri-disciplinary internal structure and increasing systemicity to be crossed by students as they advance in their studies; it is then presumably up to the student himself to form his own personal inter-disciplinary links. The systemic multilevel co-ordination of educational structures beyond teleological inter-disciplinarity (mainly on the scientific-technical side of the system) and some limited experiments in normative inter-disciplinarity has hardly

[7] Harold A. Linstone, 'A University for the Postindustrial Society', *Technological Forecasting*, Vol. 1, No. 3 (March 1970).

been considered so far in the discussions around university reform – perhaps because a clear view of the new purpose of the university is still lacking. In this section, a possible trans-disciplinary structure for the university will be outlined which the author has tried to develop with a view toward the future of the Massachusetts Institute of Technology.[8]

The basic structure of the trans-disciplinary university may be conceived as being built essentially on the feedback interaction between three types of units, all three of which incorporate their appropriate version of the unified education/research/service function:

Systems design laboratories (in particular, socio-technological systems design laboratories), bringing together elements of the physical and the social sciences, engineering and management, the life sciences and the humanities, law and policy sciences. Their tasks will not be sharply defined, but rather broad areas will be assigned to them, such as 'Ecological Systems in Natural Environments', 'Ecological Systems in Man-Made Environments', 'Information and Communication Systems', 'Transportation/Communication Systems', 'Public Health Systems', 'Systems of Urban Living', 'Educational Systems', and the like. These broad areas will, and should, overlap. Apart from designing and engineering specific systems, these laboratories will also have the task of long-range forecasting, identifying aspects and boundaries of systems emerging from the simulation of complex dynamic situations. They will be responsible for exploratory and experimental systems building at smaller scale, and they will provide opportunities for a through-flow of professionals for their self-renewal.

Function-oriented departments, taking an outcome-oriented look at the functions technology performs in societal systems, and dealing flexibly with a variance of specific technologies which all might contribute to the same function. Examples of such functions are 'Housing', 'Urban Transportation', 'Power Generation and Transmission', 'Automation and Process Control', 'Educational Technology', 'Telecommunication', 'Information Technology', 'Food Production and Distribution', etc. These functions are more clearly defined and constitute more stable 'modules' than the socio-technological systems of which they are facets. They constitute need categories which elicit the response of different technological options. Thinking in these categories implies breaking out of the linearity of specific technological development lines, and keeping the view open into a longer-range future. Education in the frame-

[8] Erich Jantsch, *Integrative Planning for the 'Joint Systems' of Society and Technology – the Emerging Role of the University*, Alfred P. Sloan School of Management, Massachusetts Institute of Technology, Cambridge, May 1969; substantial extracts have been published under the same title in *Ekistics*, Vol. 28, No. 168 (November 1969).

work of these systemic functions in society will become ever more relevant, with industry increasingly adopting a corresponding organisational framework.[9] Apart from developing technological options, which come under the heading of these functions, these departments will also emphasise systems analysis of the effects and side-effects of selecting specific technologies for satisfying needs in these areas, forecasting which will be more properly technological forecasting in its broad connotation, and assessment of the 'systems effectiveness' of technologies in the context of societal systems.

Discipline-oriented departments of a more familiar type, but with a somewhat different scope, comparatively smaller and more sharply focussed on the inter-disciplinary potential (or 'valency') of the disciplines. These departments will be mainly set up in the basic scientific disciplines at the empirical level of the education/innovation system and in the structural sciences, including such new fields as computer science.

The three layers of organisational structure focus on the inter-disciplinary co-ordination of the purposeful/normative, normative/pragmatic, and pragmatic/empirical levels of the education/innovation system.[10] The accent is here on the *links between pairs of systems levels* – in other words on *inter-disciplinary organising principles and methods* – rather than on the substance, the accumulated knowledge at the systems levels. *Figure 16.1* shows schematically how the structures of the trans-disciplinary university relate to the levels of the education/innovation system.

Unlike present university structures, focussing to an excessive degree on knowledge *per se* and (in the technological disciplines) on 'know-how', the function-oriented departments will emphasise 'know-what' – the quality which Norbert Wiener already put before 'know-how' – and the systems design laboratories the dynamic 'know-where-to', both of which are prerequisites for our ambitions to actively shape our future. The discipline-oriented departments on their side will make a new and conscious approach to 'know-why' rather than 'know-how', emphasising the investigation of basic potentials and limitations for the design of systems, in particular the 'joint systems' of society and technology. This approach may be expected to give an entirely new focus to the life sciences, concerned primarily with the feedback interaction between man and environment.

The feedback interaction between the three types of structural units in the trans-disciplinary university is sketched in *Table 16.1*.

We may then envisage a university in which some students go through discipline- and function-oriented departments only, and others go through all three types of structural units. As the latter proceed from undergraduate

[9] See Chapters 8 and 10 of this volume.
[10] See Chapter 15 of this volume, in particular *Figure 15.1*.

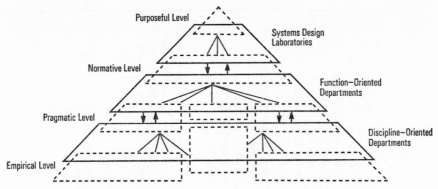

Fig 16.1 Transdisciplinary University Structure. The three types of structural units (full lines) focus on the interdisciplinary links between the four levels of the education/innovation system (dotted lines).

to graduate and doctoral work, they will shift the emphasis of their studies from discipline- and function-oriented departments more and more to the systems design laboratories, at the same time getting increasingly involved with purposeful work in technology and actual socio-technological systems design and engineering, which will become a full-time (and paid) engagement during the doctoral work. Work phases and 'absorptive' phases may alternate, with the need for theoretical learning being enhanced and guided by work. In essence, students will not go through these structural types in sequence, but interact with them simultaneously during their studies.

Such a university will turn out people with a widely varying education, from specialist-scientists over mission- and function-oriented scientists and engineers to full-scale socio-technological systems engineers. The systems design laboratories will also play an important role in the continuous education of professionals, who probably will come back to the university in much greater numbers than today.

One may believe that the outlined three-level structure will give the education function greatly increased flexibility in many respects – for specialised as well as broad (but not superficial) education, for changing tracks, for participation in various actual projects and in various qualities, for combining student and adult education, for stimulating leadership and professionalism, for education geared to various types of careers in the public and private sectors.

An important aspect concerns new dimensions in learning which may be opened up by the change from receiving training to doing useful work. With the university structure outlined here, education will take on more and more the form of self-education, and only part of it with the help of 'teachers'. A student working in a systems design laboratory will be able to

Types of units	Education	Research	Service
Systems Design Laboratories	Socio-technological systems engineers	Integrative planning and design for 'joint systems' of society and technology	'Know-where-to' through inventive contributions to public policy planning and to the active development of new socio-technological structures
Function-Oriented Departments	Stationary engineers (oriented toward functions and missions for technology, not toward engineering skills or specific technologies)	Strategic planning and development of alternatives (including innovative technological research) in areas defined by functions of technology in a socio-technological systems context	'Know-what' through providing strategic impulses to the development and introduction of technology into systems of society
Discipline-Oriented Departments	Specialist scientists	Research at the fundamental level, and development of theory	'Know-why' through clarification of the logic principles, basic potentials, and limitations inherent in empirical science

Table 16.1. The Pattern of Focal Activities in the Trans-disciplinary University*

* The higher-level activities in this scheme are always carried out through feedback interaction with the lower-placed activities in the vertical columns. All activities are horizontally integrated over the university functions of education, research, and service.

judge for himself what working and learning experience he needs from the function- and discipline-oriented departments, to which he will go back part of his time. He will be able, to a relatively large extent, to work out his curriculum himself, and to set his own educational goals and priorities. Education will move away from the stereotypes of today and become increasingly self-education in an environment which provides an infinite variety of possibilities.

This will be possible, because the student's work can be judged directly from his contribution to useful work. He may, therefore, graduate and obtain higher degrees without being examined by the rigours characteristic of the university today. No grading system will be necessary to measure the development of his capabilities. He may not even write a thesis by himself, but make corresponding contributions to team work.

Providing academic careers for all three types of structural units will give immense freedom to the entrepreneurs, and may also change the traditional status system of the university. As a matter of fact, the university professor, as we know him today, may almost vanish, or become almost indistinguishable from the students and professionals, at least in the systems design laboratories and, to some extent, in the function-oriented departments. What we call faculty today, may be the entrepreneurial leaders of the systems design laboratories tomorrow, and the through-flow of younger and older people would be identified today as students moving on in their studies, and professionals moving in and out of the university in their almost continuous education.

Viewed in the light of the research function of the university, the basic form of interaction between the three types of structural units will be a translation process in both directions between the dynamic characteristics of real and 'invented' socio-technological systems, functions and missions for technology, and contributions to them from the scientific disciplines. But the most important task in this process will be the formulation of socio-technological systems engineering requirements in terms of techno-logical missions and 'building blocks'. This task will fall primarily to the systems design laboratories.

The enhanced 'know-what' will not strangle the freedom of research, but, on the contrary, give it deeper meaning. The interaction between the three structural levels of the new university may, for the first time, lead to the investigation and active *shaping of science policy* in a rational and systematic (because systemic) way, and to its *planning and implementation in a decentralised way through the university*. This is what is called in this chapter the role of the university as a political institution. It will not be easy for the university to maintain its vitality and continuously renew itself in the erosive political process. For the first time, the university will expose itself to full public criticism, and initially suffer considerable shock from the sudden loss of its protection behind the faceless mask of 'objective' science. Already the fundamental switch toward broad, horizontal thinking across established disciplines will inevitably lead to a transitory crisis period for the university which has developed its excellence by penetrating deeply into sharply defined, more or less independently pursued disciplines.[11] However, there does not seem to be any alternative if a rational, one may even say an ecological approach to science and technology is considered mandatory, as indeed it has to be in the present situation.

It is obvious that the traditional concepts of 'value-free' science and 'neutral' technology will become completely dissolved in the unified approach, as the university proceeds to inter- and trans-disciplinarity. On the other hand, also the normative and psycho-social disciplines, such as law and sociology, will lose their abstract disciplinary identity and concepts and become aspects of social systems design. Through a trans-disciplinary approach, the university will maintain its flexibility also for future situations in which there may be less emphasis on scientific/technical aspects of social systems design, and more on human and psycho-social development. Some people expect such a shift in emphasis to become significant before the end of the century. A more short-range effect of the trans-disciplinary university may be renewed 'faith' in science and technology and a reversal of the current trend of decreasing interest of students in the scientific/ technical side of the education system.

The generalised axiomatics of the trans-disciplinary university, as it is currently shaping up in a variety of inter-disciplinary experiments, develops around what Dubos[12] calls the 'science of humanity', the science of man's total living experience. The trans-disciplinary approach thus finding its central theme, which may be understood as the new 'universitas', will be humanity-oriented. It will give the university the flexibility to abandon linear organising principles, such as the current direction and momentum of technology and its supporting sciences.

For the proper study of man, as Carey[13] sees it, 'we need something capable of shaping science goals and strategies with depth and range and visibility . . . a centre for examining the interaction of science with higher education, social change, international co-operation, technological development, and economic growth. It would be a centre to examine the mix of national investment in science and technology, to assess the quality and social returns of the investment, to identify opportunities and imbalances, to formulate models for investment that are addressed rationally to the variety of needs that we face – in short, . . . a start toward indicative planning of the uses of science and technology.'

[11] H. Guyford Stever, 'Trends of Research in Universities', in *Proceedings of the Symposium on National R&D for the 1970's*, National Security Industrial Association, Washington, D.C., 1967.
[12] René Dubos, *So Human an Animal*, Scribner, New York 1968.
[13] William D. Carey, 'Toward the Proper Study of Man', *Technology Review* (MIT), March 1969.

It is inconceivable that this task be carried out without bringing the full potential of scientific knowledge and ideas, in other words also the potential of the university, into the planning and design process. The university has to become a basic unit in a decentralised, pluralistic process of shaping the national – and, beyond that, a future global – science policy. It has to contribute to the development of a common policy for society at large, participate in the competitive process of formulating strategies, but be fully responsible for its own tactics which include the support of basic science and the development of technological skills.

The institution envisaged by Carey, may be set up as an inter-university organisation, which may become the 'melting pot' and the centre for synthesis of a group of major universities. It would provide a 'strategic antenna' oriented toward society's values as well as toward the future and maintain the dialogue with the educated public. It would force government to formulate an overall policy and it would stimulate contributions from the universities backing the institution. It would guide socio-technological systems design and engineering by giving it the proper framework.

The university will have to maintain close connection with many organisation elements of society, with government at all jurisdictional levels, with research institutes, and with industry. It will stimulate and maintain the information flow in the triangle government–industry–university, and interact actively within this triangle in the planning for society at large. Such interactions may include consortia, joint-ventures, prime and subcontracting, consultancies, etc., with government, industry, other universities, and research institutes. The systems design laboratories in the trans-disciplinary university will, in many instances, lead this process by developing innovative design proposals.

Conceivably, the university will also provide methodological aid to both government and industry, possibly through broad horizontal institutes, resembling the recently established 'Institute for the Future' in the United States, whose first research centre lives in symbiosis with Wesleyan University.

The task of turning the university from a passive servant of various elements of society and of individual and even egoistic ambitions of the members of its community into an active institution in the process of planning for society implies profound change in purpose, thought, institutional and individual behaviour. It will give the university freedom, dignity, and significance – qualities which have become grossly distorted in a process in which the university is used, but is not expected and not permitted to participate actively. The thorny way to an inter- and trans-disciplinary university has been outlined in this chapter as the way to a new and active role of the university in society.

Selected Further Readings

Long-Range and Systemic Planning:

C. West Churchman, *Challenge to Reason*, McGraw-Hill, New York, 1968.

—— *The Systems Approach*, Delacorte Press, New York, 1968.

—— *The Design of Inquiring Systems*, Basic Books, New York, 1971.

René Dubos, *Reason Awake: Science for Man*, Columbia University Press, New York, 1970.

Dennis Gabor, *Inventing the Future*, Secker & Warburg, London, 1963, Penguin Paperback, London, Harmondsworth, 1964, and Knopf, New York, 1964.

Erich Jantsch (ed.), *Perspectives of Planning*, OECD, Paris, 1969.

Donald Michael, *The Unprepared Society: Planning for a Precarious Future*, Basic Books, New York, 1968, and Harper & Row Colophon paperbacks, 1969.

Geoffrey Vickers, *The Art of Judgement*, Basic Books, New York, 1965.

—— *Towards a Sociology of Management*, Basic Books, New York, 1967.

—— *Value Systems and Social Process*, Tavistock Publications, London, 1968, Penguin, Harmondsworth, 1970, and Basic Books, New York, 1968.

—— *Freedom in a Rocking Boat: Changing Values in an Unstable Society*, Allen Lane, The Penguin Press, London, 1970.

Corporate Planning:

Russell L. Ackoff, *A Concept of Corporate Planning*, Wiley-Interscience (John Wiley & Sons), New York, 1970.

George A. Steiner, *Top Management Planning*, Macmillan, New York, 1969.

Systems:

Fred E. Emery (ed.), *Systems Thinking*, Penguin, Harmondsworth, 1969.

T.P.A.S.F.—R

Jay W. Forrester, *Industrial Dynamics*, M.I.T. Press, Cambridge, Mass., 1961.

———   *Urban Dynamics*, M.I.T. Press, Cambridge, Mass., 1969.

———   *World Dynamics*, Wright-Allen, Cambridge, Mass., 1971.

Mihajlo D. Mesarović, D. Macko, and Y. Takahara, *Theory of Hierarchical, Multilevel, Systems*, Academic Press, New York and London, 1970.

Forecasting:

Robert U. Ayres, *Technological Forecasting and Long-Range Planning*, McGraw-Hill, New York, 1969.

James R. Bright (ed.), *Technological Forecasting for Industry and Government*, Prentice-Hall, Englewood Cliffs, N.J., 1968.

Olaf Helmer, *Social Technology*, Basic Books, New York, 1966.

Erich Jantsch, *Technological Forecasting in Perspective*, OECD, Paris, 1967.

Bertrand de Jouvenel, *The Art of Conjecture*, Basic Books, New York, 1967.

Gordon Wills, B. Ashton, and B. Taylor (eds.), *Technological Forecasting and Corporate Strategy*, Bradford University Press in association with Crosby Lockwood, London, 1969, and Elsevier, New York, 1969.

Michael Young (ed.), *Forecasting and the Social Sciences*, Heinemann, London, 1968.

Technology and the 'Whole World' Perspective:

Victor C. Ferkiss, *Technological Man: The Myth and the Reality*, Braziller, New York, 1969, and Heinemann, London, 1969.

Jay W. Forrester, *World Dynamics* (see under 'Systems' above).

Massachusetts Institute of Technology, *Man's Impact on the Global Environment: Assessment and Recommendations for Action*, Report of the Study on Critical Environmental Problems (SCEP), M.I.T. Press, Cambridge, Mass., 1970.

Aurelio Peccei, *The Chasm Ahead*, Macmillan, New York and London, 1969.

Geoffrey Vickers, *Freedom in a Rocking Boat: Changing Values in an Unstable Society* (see under 'Long-Range and Systemic Planning' above).

New Corporate Roles:

(Daedalus), *Perspectives on Business*, Daedalus, Winter 1969.

Clarence C. Walton (ed.), *Business and Social Progress: Views of Two Generations of Executives*, Praeger, New York, 1970.

Journals:

Futures, quarterly, since September 1968, IPC Press, Guildford, Surrey.

Innovation, ten issues per year, since May 1969, Technology Information Group, New York.

Long Range Planning, monthly, since September 1968, Pergamon Press, Oxford.

Technological Forecasting and Social Change, quarterly, since June 1969, American Elsevier Publishing Co., New York.

Name Index

Subject Index